NO GRID SURVIVAL PROJECTS

BIBLE

Secure Your Home, Power, Water and Food Supply
to Thrive Through Any Crisis & Disaster.
The Only Guide You Will Ever Need for Self-Sufficiency & Independence

by

Jonathan A. Hugg

Table of Contents

Disclaimer & Legal Notice

The information presented in this document is provided for educational and entertainment purposes only. While every effort has been made to ensure accuracy, currency, and reliability, no guarantees are made as to the completeness or correctness of the content. This document is not intended to offer legal, financial, medical, or any other form of professional advice. Readers are encouraged to consult with qualified professionals before applying any techniques or practices discussed in this document.

By accessing this material, the reader acknowledges and agrees that the author will not be held responsible for any direct, indirect, incidental, or consequential damages resulting from the use of the information provided. This includes any losses that may arise from errors, omissions, or inaccuracies in the content.

Copyright © 2024

Welcome Aboard

Creating this guide has been a journey fueled by dedication and a passion for providing you with the essential knowledge and hands-on skills needed to thrive in a world where self-reliance is key. Your support plays a vital role in driving us forward, and helping us empower others to embrace off-grid living and survival. Thank you for being such an important part of this mission!

Before you dive into the projects, let's go over a couple of important steps:

1) Access Your Bonus Content: Be sure to take full advantage of the extra resources included with this book. These materials are designed to enhance your understanding and help you successfully implement the off-grid survival projects. To access your bonus content, simply head to the section at the end of the book titled "Retrieve Your Bonus Content."

2) Share Your Thoughts: We'd love to hear how excited you are about having this book in your hands! Your feedback helps others who are looking for practical ways to live off the grid. You can choose how you'd like to share your thoughts:

- Option A: Record a short video and write a few sentences

- Option B: If you prefer not to be on camera, no problem! You can snap a few pictures of the book or simply share a brief message.

Scan the QR code below to share your excitement!

Module A | Planning and Setting Up Your Off-Grid Home

1. General Principles and Guidelines to Build an Off-Grid Home

Living off the grid offers freedom, sustainability, and a return to simpler, more intentional living. But before you dive into building your dream off-grid home, it's essential to understand the principles that will help you create a space that is efficient, resilient, and self-sustaining. In this section, we'll explore the different types of off-grid homes, sustainable building materials, insulation techniques, site layout strategies, and much more to ensure that your off-grid home is built for long-term success.

Types of Off-Grid Homes

When it comes to off-grid living, there is no one-size-fits-all solution. The type of home you choose will depend on your location, climate, budget, and personal preferences. Here are a few common types of off-grid homes that offer flexibility and sustainability:

Cob Houses

Cob houses are constructed from a mixture of earth, straw, and water. This building technique has been used for centuries and is known for its thermal mass properties, meaning it can absorb and store heat, keeping your home warm in the winter and cool in the summer. Cob is highly durable and environmentally friendly, using locally sourced materials with little environmental impact. These homes are often sculptural in appearance, making them visually unique and naturally insulated.

Earthships

An earthship is a self-sustaining home made primarily of natural and recycled materials like tires filled with earth, glass bottles, and aluminum cans. Earthships are designed to be highly energy-efficient and often include solar power, water harvesting systems, and indoor gardening areas. Their thick walls provide excellent insulation, and they are often built partially underground to take advantage of the earth's natural temperature regulation. These homes are ideal for those looking for an entirely off-grid lifestyle with minimal dependence on external resources.

Tiny Homes

Tiny homes have gained popularity due to their low cost and small environmental footprint. These homes typically range from 100 to 400 square feet and can be constructed on wheels or a permanent foundation. Tiny homes can be customized with solar panels, rainwater harvesting systems, and composting toilets, making them ideal for off-grid living. Their compact size also means they require fewer resources to heat and cool, making them energy-efficient.

Yurts

Yurts are circular, tent-like structures that have been used for thousands of years by nomadic peoples in Central Asia. Modern yurts are often made with a wooden frame and canvas or fabric covering, though more permanent versions may incorporate insulation and durable materials like wood or metal. Yurts are lightweight, relatively easy to set up, and can be customized with modern off-grid systems like solar power and wood stoves. They offer a quick and inexpensive way to set up an off-grid home in remote locations.

Shipping Containers

Shipping container homes are made from repurposed steel containers, offering a sturdy and weather-resistant structure. These homes are highly customizable, allowing you to stack or combine containers to create different layouts. Shipping containers are incredibly durable, withstanding harsh weather conditions, and can be easily insulated and outfitted with off-grid systems. They are ideal for those looking for a modern, industrial aesthetic combined with sustainability.

Building Materials

Selecting the right building materials is one of the most important decisions when constructing an off-grid home. The materials you use should be sustainable, durable, and ideally locally sourced to reduce the environmental footprint. Here are some of the best options for building off-grid:

Adobe

Adobe bricks are made from a mixture of clay, sand, and straw, and they are dried in the sun rather than fired in a kiln. Adobe homes are incredibly durable and have excellent thermal mass properties, meaning they help regulate indoor temperatures. Adobe is particularly well-suited for hot, dry climates as the thick walls retain coolness during the day and warmth at night. Building with adobe is labor-intensive but provides a long-lasting and sustainable solution.

Straw Bale

Straw bale construction involves using tightly packed bales of straw as walls, which are then covered in plaster or clay. Straw bale homes are extremely energy-efficient, providing excellent insulation that keeps your home cool in the summer and warm in the winter. Straw is a renewable resource, making it an environmentally friendly building material. Plus, straw bale construction is surprisingly strong and can stand up to extreme weather conditions.

Reclaimed Wood

Reclaimed wood is sourced from old buildings, barns, or even pallets, offering a sustainable alternative to freshly cut lumber. Using reclaimed wood reduces the demand for new materials and gives a rustic, weathered aesthetic to your home. It's an excellent choice for framing, flooring, and interior finishes. However, it's important to ensure that the wood is treated and free of pests before using it in construction.

Rammed Earth

Rammed earth is a technique where natural earth is compressed into molds to form thick walls. Like cob and adobe, rammed earth has high thermal mass, making it an ideal choice for passive heating and cooling. It is a low-cost, low-maintenance option that blends seamlessly into the natural environment. Rammed earth homes are often incredibly durable and can last for centuries if properly maintained.

Metal and Steel

For more modern off-grid homes, metal and steel can provide strength and durability. Steel-framed homes or those made from shipping containers are resistant to fire, pests, and extreme weather. However, they do require more insulation since metal conducts heat, making it less energy-efficient without proper treatment.

When selecting materials, consider their availability, environmental impact, and the climate in which you're building. Materials with high thermal mass, like cob or adobe, are excellent for areas with extreme temperature fluctuations, while wood or straw may be better suited for temperate regions.

Insulation and Passive Solar Design

A key factor in building an off-grid home is energy efficiency, and insulation is crucial to achieving that goal. Proper insulation helps retain heat in the winter and keep your home cool in the summer, reducing your reliance on external energy sources.

Natural Insulation Materials

When building an off-grid home, using natural insulation materials like wool, straw, or hemp can provide excellent thermal protection while remaining environmentally friendly. Wool, for example, is naturally fire-resistant, breathable, and renewable. Straw bales, as mentioned earlier, are also highly insulative and cost-effective.

Recycled Insulation

Recycled materials like denim, cotton, and cellulose (made from recycled paper) are becoming increasingly popular as sustainable insulation options. These materials are easy to install, non-toxic, and offer good thermal resistance.

Passive Solar Design

Passive solar design involves positioning your home to take advantage of the sun's natural energy. By aligning windows and living spaces to the south (in the northern hemisphere), you can capture maximum sunlight during the winter, reducing the need for artificial heating. Thermal mass materials like concrete or adobe can absorb and store this heat, slowly releasing it throughout the night.

Ventilation for Temperature Control

Proper ventilation is critical to maintaining a comfortable indoor climate. Designing windows and vents to allow for natural cross-ventilation can help cool your home during the summer without the need for air conditioning. Roof overhangs and shading can prevent excessive heat buildup, while properly sealing your home will keep cold air out during the winter months.

Orientation and Site Layout

The orientation and layout of your off-grid home are just as important as the materials you use. Proper planning can significantly improve energy efficiency, water management, and overall comfort.

Positioning for Solar Gain

In off-grid homes, solar power is often a primary energy source. Position your home to maximize sunlight exposure, with the longest walls facing south (in the northern hemisphere) to capture winter sun. This also provides natural light, reducing the need for artificial lighting.

Wind and Weather Protection

Consider the prevailing winds and weather patterns in your area. Placing your home in a sheltered location, such as behind a natural windbreak or slope, can help protect it from strong winds, storms, and extreme weather.

Site Layout for Water and Drainage

Ensure your home is positioned on higher ground or a slight slope to prevent flooding. Proper water drainage is critical to avoid foundation damage and manage stormwater. Incorporating swales or ditches around your property can help guide rainwater to where it's needed, such as your garden or water storage system.

Maximizing Outdoor Space

Design your site layout to include gardens, orchards, or livestock areas. This ensures your home isn't just a shelter but part of a self-sustaining ecosystem. Consider incorporating permaculture principles to create a balanced, productive environment.

Water Drainage and Flood Prevention

Off-grid living often means relying on natural water sources or rainwater collection, so planning for proper drainage and flood prevention is essential.

Grading and Sloping

Grading your land to create a gentle slope away from your home ensures that water flows away from your foundation. This helps prevent flooding and water damage, which is especially important in areas prone to heavy rainfall or seasonal flooding.

Swales and Berms

Swales (shallow ditches) and berms (raised soil) are simple but effective ways to manage water on your property. Swales can capture rainwater and direct it toward your garden, while berms help prevent erosion and flooding.

Rainwater Harvesting

Installing rain barrels or larger cisterns to capture rainwater is an excellent way to supplement your water

supply while managing runoff. This can be combined with a gutter system to ensure that water is directed to the right storage locations.

Landscaping for Self-Sufficiency

Landscaping goes beyond aesthetics in an off-grid home—it plays a crucial role in food production, energy efficiency, and privacy.

Food Production

Design your landscaping to include raised garden beds, greenhouses, or even a food forest. This ensures that your land is working for you, providing fruits, vegetables, and herbs throughout the year.

Windbreaks and Privacy

Planting trees or installing natural barriers like hedges can provide wind protection, reduce energy loss, and offer privacy. Windbreaks help prevent soil erosion and can create microclimates that are more suitable for growing crops.

Water Conservation

Use landscaping techniques like mulching, drip irrigation, and planting drought-tolerant species to reduce water usage. This is especially important in areas with limited water resources.

Ventilation and Air Quality

Maintaining good air quality in an off-grid home is critical for health and comfort, especially in homes built with natural materials.

Natural Ventilation

Design your home to include plenty of windows and vents to promote cross-ventilation. This allows fresh air to circulate and prevents the buildup of humidity, which can lead to mold and mildew in humid environments.

Heat Recovery Ventilators

In colder climates, heat recovery ventilators (HRVs) can be used to improve indoor air quality while retaining heat. These systems bring in fresh air and expel stale air, but they capture the heat from the outgoing air to reduce energy loss.

Indoor Plants for Air Purification

Including houseplants like aloe vera, spider plants, or peace lilies can help purify indoor air by absorbing pollutants and releasing oxygen. This not only improves air quality but adds a natural touch to your living space.

By understanding these principles and guidelines, you'll be well-equipped to build an off-grid home that is both efficient and sustainable. Each decision you make—whether it's the type of home, the materials you use, or the way you design your layout—can contribute to a more self-sufficient and resilient lifestyle.

2. Budgeting and Resource Planning for Your Off-Grid Home

Building an off-grid home is a major commitment, not just in terms of lifestyle but also in financial planning. Proper budgeting is critical to ensure that your project remains feasible, sustainable, and adaptable to your evolving needs. In this section, we'll walk through how to estimate the costs for land, materials, and labor, while also considering ways to reduce costs through resourceful building practices. We'll also explore planning for the future with modular home designs and long-term sustainability strategies.

Budgeting for an Off-Grid Home Build: Estimating Costs for Land, Materials, and Labor

Creating a detailed budget is the first crucial step in building your off-grid home. Unlike traditional homes, off-grid builds come with unique financial considerations, such as energy systems, water sourcing, and sustainable materials. Here's how you can break down the key components of your budget.

1. Land Costs

The cost of land is often the first major expense, and it can vary significantly depending on location, accessibility, size, and natural resources. When budgeting for land:

- Remote Land vs. Proximity to Towns: Remote land tends to be more affordable, but it may also come with added costs like road building, extended utility connections (if applicable), or additional transportation needs.

- Water and Resource Availability: Land with access to natural water sources like rivers, springs, or underground wells may cost more but could save you significant money in the long term by reducing the need for expensive water collection systems.

- Zoning and Legal Costs: Consider the costs of permits, zoning regulations, and land inspections to ensure that you can legally build off-grid on your chosen plot.

- Initial Land Preparation: Clearing land, leveling for foundations, and installing basic infrastructure like driveways or access roads may incur additional costs. Budget for heavy machinery or professional help if the land requires significant alteration.

2. Materials Costs

One of the key principles of off-grid living is sustainability, which extends to the materials used in construction. Material costs will vary depending on the type of home you are building (e.g., cob, earthship, tiny home, or a more conventional structure). However, all builds will have some common material costs:

- Foundation Materials: Concrete, rammed earth, or recycled materials for the foundation. Depending on your design, this could be one of the more expensive parts of the build.

- Structural Materials: Timber, steel, or alternative natural materials like straw bales or adobe. The cost will depend on the sustainability of the materials, the availability in your region, and whether you're sourcing locally or having materials shipped.

- Energy Systems: Solar panels, wind turbines, battery banks, or generators may require an upfront investment, but they are essential for energy independence. These systems should be factored into your materials budget, as they will have a significant impact on the overall cost.

- Water and Plumbing Systems: The cost of water storage tanks, filtration systems, pumps, and plumbing materials will also contribute to your budget. Wells, rainwater harvesting systems, or greywater recycling systems can be cost-effective in the long run but may have a high initial outlay.

3. Labor Costs

Labor can be a major cost in any home build. One way to reduce this expense is through DIY building, but it's important to understand the time and effort required. Here are some considerations:

- DIY vs. Hiring Professionals: DIY labor can significantly reduce costs, but it may require acquiring new skills, buying specialized tools, and dedicating a significant amount of time. For highly technical aspects like electrical work or plumbing, hiring a professional may save you time and ensure the job is done correctly and safely.

- Skilled Tradespeople: Depending on the complexity of your build, you may need to hire specialists such as electricians, plumbers, or masons. Budget for these services, even if you plan to do the majority of the work yourself.

- Project Management: If you are not managing the project yourself, hiring a project manager or contractor to oversee the build can be an additional cost. However, this can save time and reduce the likelihood of mistakes or delays.

By carefully planning your budget across these categories, you can get a realistic picture of how much your off-grid home will cost and where you may need to allocate more resources or cut back.

Self-Sufficient Building: Using Reclaimed, Recycled, or Locally Sourced Materials

One of the most effective ways to reduce the cost of building an off-grid home is to use reclaimed, recycled, or locally sourced materials. Not only does this lower your expenses, but it also significantly reduces the environmental impact of your build.

1. Reclaimed Materials

Reclaimed materials are sourced from old buildings, barns, or other structures that have been dismantled. These materials are often still in excellent condition and can add character and sustainability to your home.

- Reclaimed Wood: Using wood from old barns or construction sites is an excellent way to cut down on costs while incorporating high-quality, aged timber into your build. It's perfect for framing, flooring, or decorative elements.

- Salvaged Doors and Windows: Architectural salvage yards are excellent places to find doors, windows, and fixtures that can be repurposed for your off-grid home. Not only do they reduce waste, but they also offer unique design opportunities.

- Recycled Metal: Metal roofing or structural elements can often be salvaged from old industrial buildings. Steel and other metals are highly durable and can be repurposed without losing strength or integrity.

2. Recycled Materials

Incorporating recycled materials into your build is both cost-effective and eco-friendly. Many off-grid homes make use of recycled products to reduce waste and reliance on new materials.

- Tires for Earthships: Old tires, filled with compacted earth, form the foundation of many earthship homes. These walls are durable, provide excellent thermal mass, and make use of materials that would otherwise end up in landfills.

- Recycled Glass: Glass bottles can be used creatively for insulation or as decorative elements in walls. This is not only an inexpensive material but also a great way to incorporate artistic touches into your home design.

- Shipping Containers: Repurposing shipping containers for housing is growing in popularity due to their availability and structural integrity. Shipping containers can be an affordable alternative to traditional building materials, providing a strong framework for an off-grid home.

3. Locally Sourced Materials

Using materials sourced from your local environment reduces transportation costs and supports sustainable building practices. Depending on your location, certain natural materials may be readily available.

- Stone and Earth: Many off-grid homes use local stone for foundations, walls, or decorative features. In areas with abundant clay, adobe or cob construction is a sustainable option that can significantly cut material costs.

- Straw Bales: Straw is a renewable resource that can be found locally in many rural areas. Straw bales are not only affordable but also provide excellent insulation for both hot and cold climates.

- Bamboo: In tropical or temperate regions, bamboo is a fast-growing, sustainable material that can be used for framing or as a structural element in your home. It's strong, lightweight, and naturally resistant to pests.

By choosing reclaimed, recycled, or locally sourced materials, you can save thousands of dollars while also building a home that is more in harmony with the environment.

Time and Labor Considerations: DIY vs. Professional Help

Building an off-grid home can be a rewarding DIY project, but it requires a realistic assessment of how much time, effort, and skill you can dedicate to the build. Here's how to balance doing the work yourself with hiring professionals when needed.

1. DIY Building

Many off-grid homeowners choose to build their homes themselves to save money and gain a deeper connection to the land. However, a DIY approach comes with challenges, especially for those who don't have experience in construction.

- Skill Acquisition: Depending on the complexity of your design, you may need to learn new skills like carpentry, masonry, plumbing, or electrical work. This learning curve can slow the build and potentially lead to mistakes that may need correction later.

- Time Commitment: DIY builds take time, often stretching into months or even years. Consider whether you have the time to dedicate to a large-scale project while balancing other life responsibilities like work and family.

- Cost of Tools and Equipment: While DIY building reduces labor costs, you will need to invest in tools and equipment. Renting or purchasing specialized tools (e.g., for cutting timber or welding metal) can add to your budget.

2. Hiring Professionals

For more technical aspects of the build, it's often worth hiring professionals to ensure that the job is done safely and up to code.

- Plumbing and Electrical Work: Electrical and plumbing systems are critical to the safety and functionality of your home. Mistakes in these areas can lead to serious issues, such as fires or water damage, so it's often best to leave these tasks to licensed professionals.

- Structural Elements: If you're building with unconventional materials like cob or rammed earth, consider hiring an expert to ensure the structural integrity of your home.

- Saving Time with Contractors: Hiring a contractor can expedite the build process, ensuring that your home is completed on time. This is especially useful if you have a strict deadline or need to finish the home before winter sets in.

3. Hybrid Approach

Many off-grid homeowners take a hybrid approach, tackling simpler tasks like framing, roofing, and interior work themselves, while hiring professionals for more complex systems. This allows you to save money while still ensuring that the critical elements are handled by experts.

Planning for Future Expansion

One of the benefits of building your own off-grid home is the ability to plan for future growth and changes. Whether you want to add more space for a growing family, incorporate new technologies, or adapt to environmental changes, designing with expansion in mind is a smart move.

1. Modular Home Design

Modular design is a flexible approach that allows you to add new sections to your home as needed. This is particularly useful for off-grid homes, where you may want to start small and gradually expand your living space.

- Expandable Structures: By designing your home with modular units, you can easily add new rooms, outbuildings, or greenhouses over time. This allows you to spread out the cost of construction and adapt your home to changing needs.

- Pre-planned Utility Lines: If you plan to expand in the future, ensure that your electrical, plumbing, and HVAC systems are designed to accommodate additional structures. Running utility lines early will save you time and money down the road.

2. Adaptability for Changing Needs

Life changes, and so should your home. As you plan your off-grid build, consider how your needs might evolve over the years.

- Family Growth: If you plan to expand your family, make sure your design includes enough space to accommodate new members. Even if you start with a small home, having extra land available for future expansion can make your home more adaptable.

- Technological Upgrades: Technology for off-grid living is constantly evolving. Whether it's upgrading your solar panels or adding new water filtration systems, design your home to easily integrate future technologies.

- Environmental Adaptations: Climate change may alter the conditions in your area over time. Consider how you might need to adapt your home to new weather patterns, such as adding insulation, reinforcing structures for extreme weather, or improving water conservation systems.

3. Long-Term Sustainability

Building for sustainability means considering how your home will function not just for the next few years, but for decades. Here's how to ensure your off-grid home remains resilient over the long term.

- Energy Independence: Designing your home with energy independence in mind means incorporating renewable energy sources like solar, wind, or hydro power. Ensure that your systems are scalable so that you can expand your energy production as needed.

- Resource Availability: Consider the long-term availability of resources like water, firewood, and food. By incorporating permaculture principles or building a rainwater harvesting system, you can ensure that your home remains self-sufficient even in times of scarcity.

- Durability of Materials: Choose building materials that are designed to last. Stone, steel, and properly treated wood can withstand the test of time, while materials like adobe or rammed earth provide excellent thermal mass and durability.

By carefully planning your budget and considering future expansion, you can build an off-grid home that meets your immediate needs while also providing flexibility for the future. Whether you take a DIY approach, hire professionals, or combine both, understanding the financial and labor considerations will help you create a home that is both functional and sustainable for years to come.

3. Considerations About the Location for Your Off-Grid Home

Choosing the right location for your off-grid home is one of the most critical decisions you'll make on your journey toward self-sufficiency. The location you select will impact everything from the availability of natural resources, to how sustainable your energy systems will be, to the legal hurdles you might face. It will also determine how remote and independent you can be while maintaining access to necessary resources like food, emergency services, and community support.

In this section, we will explore four key factors to consider when selecting your off-grid location: access to natural resources, the legal landscape and regulations, environmental challenges, and accessibility to nearby towns and services. Making an informed decision in each of these areas will ensure that your off-grid home is both practical and sustainable in the long run.

Access to Natural Resources & Climate

The right location can make or break your off-grid project. It determines your access to the basic elements of self-sufficient living: water, energy, and food. When choosing a location, it's important to weigh the following factors:

1. Access to Natural Resources

For any off-grid home, access to essential natural resources like water, sunlight, and land for food production is fundamental. The abundance or scarcity of these resources will dictate many of the decisions you make when planning your off-grid lifestyle.

Water Sources

One of the most critical considerations is access to water. This is especially important if you're planning to live entirely off-grid without reliance on municipal systems. Look for land that has a natural water source, such as a stream, river, lake, or natural spring. If none of these are available, groundwater sources are another option, but drilling a well can be a costly endeavor, so ensure the land has a high water table.

Rainwater collection is another option for water independence. However, the location's rainfall patterns must be studied carefully. Areas with seasonal droughts or inconsistent rainfall may require substantial storage systems to ensure water availability year-round. In contrast, regions with high rainfall may also require flood prevention measures to protect your home and crops.

Solar Exposure

If you're planning on using solar energy, adequate sunlight is a necessity. Choose a location that has consistent sun exposure, especially in the winter when the days are shorter and your energy needs might be higher. South-facing slopes (in the Northern Hemisphere) are ideal for maximizing sunlight exposure.

Additionally, take into account any shading from nearby trees, mountains, or buildings that could reduce the amount of sunlight your panels receive. In heavily wooded areas, you may need to clear some land to ensure sufficient solar gain, though this comes at the cost of losing some of the natural beauty and environmental benefits of trees.

Wind Energy Potential

If you plan to use wind energy, the location's wind patterns are just as important as solar exposure. Open plains, coastal areas, and hilltops tend to have stronger and more consistent winds, making them ideal for wind turbines. However, in mountainous or heavily forested areas, wind energy might not be feasible due to natural obstructions.

Wind mapping tools are available that can help you estimate the wind speeds in a given area. It's essential to have consistent wind speeds of at least 10 miles per hour (16 km/h) to make wind energy a viable option.

Land for Food Production

Finally, consider the quality of the soil and available land for growing food. If you plan to raise livestock or cultivate crops, you'll need to ensure that the soil is fertile and suitable for agriculture. Conduct soil tests to determine its composition and nutrient levels. Areas with poor soil might require you to invest more in soil amendments, greenhouses, or alternative growing methods like hydroponics or aquaponics.

If you plan to forage or hunt, consider the local wildlife and plant life. Is there enough diversity to sustain you, or will you need to supplement your diet with purchased goods? Areas rich in wild edibles or game can reduce your reliance on grocery stores.

2. Climate

Climate plays a significant role in the design of your home and the systems you'll need to put in place. Different climates pose different challenges and require unique solutions for heating, cooling, water collection, and food production.

Cold Climates

In colder climates, you'll need to prioritize insulation and heating systems. Look for land that provides access to firewood if you plan to use a wood-burning stove, or consider investing in a geothermal or passive solar heating system. Snow accumulation can also be a concern, so make sure your site is accessible in the winter and that your home design can withstand heavy snow loads.

Growing food in cold climates can also be challenging. Short growing seasons may require greenhouses or hoop houses to extend the growing period. Root cellars are commonly used in these regions to store food during the winter months.

Hot and Arid Climates

If you're building in a desert or semi-arid region, water will be your primary concern. Ensure that you have access to a reliable water source, and consider implementing rainwater harvesting systems. Xeriscaping, which involves landscaping with drought-tolerant plants, can also help reduce water usage.

In hot climates, passive cooling techniques like earth-sheltered homes, thick adobe or rammed earth walls, and shade structures are essential for maintaining a comfortable indoor temperature. Solar energy is often abundant in these areas, making it an ideal location for solar power systems.

Tropical Climates

Tropical climates, while generally rich in rainfall, can pose challenges such as high humidity, intense heat, and frequent storms. Ventilation and humidity control are essential for preventing mold and maintaining indoor comfort. Building on stilts or with elevated foundations can help protect your home from flooding, while hurricane-resistant designs may be necessary in storm-prone areas.

The fertile soil and extended growing seasons in tropical areas make them ideal for permaculture and agroforestry. However, pest control may be more difficult due to the high biodiversity of insects and animals.

Understanding the Legal Landscape and Local Regulations

Before you start building your off-grid home, it's essential to understand the legal landscape and local regulations. Not all areas are zoned for off-grid living, and some regions have strict building codes that may affect your plans.

1. Zoning Laws

Zoning laws dictate how land can be used, and these regulations vary significantly depending on your location. Before purchasing land, research the zoning restrictions in your area to ensure that you can legally live off-grid.

Residential vs. Agricultural Zoning

Some areas are zoned strictly for residential or agricultural use, which can affect whether you're allowed to build an off-grid home, install renewable energy systems, or cultivate crops. Agricultural zoning typically allows for more flexibility when it comes to farming, raising livestock, and building unconventional homes, while residential zoning may come with more restrictions.

In some cases, you may be able to apply for zoning variances or conditional use permits that allow you to live off-grid, but this process can be time-consuming and costly.

Minimum Lot Size Requirements

Some counties have minimum lot size requirements, particularly in rural areas. These regulations are designed to limit urban sprawl, but they can also prevent you from purchasing small, affordable plots of land. Make sure that the land you're considering meets the minimum lot size for the type of home you want to build.

2. Building Codes and Permits

Building codes are a set of standards that regulate construction to ensure safety and compliance with local laws. Depending on your location, building codes may govern everything from the type of foundation you use to the insulation in your walls.

Alternative Building Materials

Off-grid homes often incorporate non-traditional materials like straw bales, earthbags, or cob, which may not be explicitly covered under local building codes. In some areas, you may need to get special approval or have an engineer sign off on your design to ensure that it meets safety standards.

In more progressive regions, building codes may have specific provisions for sustainable building materials and techniques, making it easier to build an off-grid home. Always check with local authorities to understand what is permitted.

Electrical and Plumbing Permits

Many counties require permits for electrical work and plumbing, even in off-grid homes. If you plan to install solar panels, wind turbines, or alternative sewage systems (like composting toilets or greywater systems), you may need special permits or inspections. Be prepared to navigate this process by hiring licensed professionals if necessary.

3. Water Rights

In some areas, water rights can be a complex legal issue. You may not automatically have the right to access groundwater or surface water on your property, even if it's located directly on your land.

Rainwater Harvesting

While rainwater harvesting is legal in most areas, some states or counties have restrictions on how much water you can collect or whether you can divert natural waterways. Be sure to understand your local water laws before implementing a rainwater collection system.

Well Drilling and Groundwater Access

Drilling a well typically requires a permit, and in some areas, you may be restricted on how much water you can extract from the groundwater table. Additionally, the quality of the groundwater can vary, so it's essential to have the water tested before committing to a well-based system.

Challenges Related to Climate, Geography, and Self-Sufficiency

Choosing the right off-grid location also means accounting for the challenges posed by your local climate and geography. These factors can affect everything from your ability to generate energy to the types of crops you can grow.

1. Climate Extremes

Living off the grid often means living closer to nature, which makes you more vulnerable to extreme weather events. Whether it's intense summer heat, freezing winters, or powerful storms, your location will dictate the systems you need to put in place to remain comfortable and safe.

Natural Disasters

Consider whether your area is prone to natural disasters like floods, hurricanes, tornadoes, or wildfires. In disaster-prone areas, you'll need to design your home to withstand these threats, which may increase construction costs. You may also need to establish an emergency plan for evacuations or supply shortages.

Growing Seasons

Your location's climate will affect your ability to grow food year-round. In northern regions with short growing seasons, you may need to invest in greenhouses or cold frames to extend your harvests. In warmer climates, you might face challenges related to drought, requiring more advanced water storage or irrigation systems.

2. Terrain and Soil Quality

The terrain and soil quality of your location are key considerations when planning for self-sufficiency.

Steep or Rocky Terrain

While steep hillsides can provide excellent opportunities for terracing or water collection, they can also make construction more difficult and expensive. Building on uneven or rocky terrain may require specialized foundations or retaining walls, adding to your overall costs.

Soil Fertility

The fertility of the soil will impact your ability to grow food and raise livestock. If the soil is poor or heavily compacted, you may need to invest in soil amendments or raised beds to cultivate crops. It's also essential to assess the risk of erosion, particularly if you're planning to farm on a slope or near a water source.

Accessibility and Transportation

While the allure of remote living is a big part of going off-grid, it's important to consider how accessible your location is, both for your own needs and for emergency services.

1. Proximity to Towns and Services

Even the most self-sufficient off-grid homesteader may occasionally need to visit a town for supplies, medical care, or social interaction. Consider how far you're willing to travel for these essentials.

Grocery Stores and Supplies

While you may be growing much of your food, there will still be times when you need to purchase staples like flour, salt, or household items. If your property is too far from a town, the cost and effort of traveling for supplies can quickly add up.

Additionally, you may need to make periodic trips to hardware stores for tools, building materials, or repairs. Having a reliable vehicle and access to a nearby town can reduce the burden of these trips.

Medical and Emergency Services

If you or a family member has a medical emergency, how long will it take for help to arrive? Consider the distance to the nearest hospital or clinic, as well as the availability of emergency services like fire trucks or ambulances. In remote areas, you may need to be more self-reliant, stocking first aid supplies and having a plan for transportation in case of an emergency.

2. Road Access and Transportation Infrastructure

The condition of the roads leading to your property is another critical consideration. Remote areas often have poorly maintained or unpaved roads, which can become impassable in bad weather.

All-Season Access

Make sure your property has access to reliable roads that are usable in all seasons. If heavy snowfall, mud, or flooding can block your access, you may need to invest in off-road vehicles or plan for longer periods of isolation during extreme weather.

Off-Road or Alternative Transportation

In very remote areas, traditional vehicles may not be sufficient for transportation. Off-road vehicles like ATVs or snowmobiles can help you navigate difficult terrain. Alternatively, you might rely on bicycles, horses, or even boats if you're located near a waterway.

Conclusion

Choosing the right location for your off-grid home is a balancing act between the resources you need, the challenges you're willing to face, and your long-term goals for self-sufficiency. By carefully considering factors like natural resource availability, legal requirements, climate, and accessibility, you can find a location that supports your off-grid lifestyle and ensures your success for years to come.

Module B | Water Independence: Securing Reliable Water Sources for Off-Grid Living

Water is the foundation of life, and when living off-grid, securing a reliable, sustainable source is crucial. This chapter will guide you through the various methods of sourcing and collecting water, providing both practical advice and detailed instructions to ensure you have enough water to meet your daily needs, even in the driest of seasons. We'll explore natural water sources, the difference between groundwater and surface water, and how to effectively harvest and store rainwater.

4. Water Sourcing Systems

When planning your off-grid water supply, the first step is determining where you can source your water. Depending on your location, you may have access to natural water sources like wells, springs, lakes, or rivers. Alternatively, you can set up systems to collect rainwater. Each of these options comes with its own set of advantages and challenges, and it's essential to understand them fully before committing to one or more solutions.

1. Natural Water Sources: Wells, Springs, Lakes, Rivers, and Rainwater Collection

For off-grid living, natural water sources are often the most reliable and self-sustaining options. If you're fortunate enough to have access to a well, spring, lake, or river, you can create a system that continuously provides fresh water without reliance on external supplies. Let's break down the common types of natural water sources you might consider:

Wells

Drilling a well is one of the most popular methods for sourcing water off-grid, especially in areas without access to natural surface water. A well taps into groundwater, which is typically found several feet or more below the earth's surface.

Steps to install a well:

- Site Selection: Before drilling, you'll need to locate a spot with a high likelihood of water. A professional hydrologist or well driller can assess your land and help you find the most productive spot.

- Permits: In many areas, you will need to obtain a permit to drill a well. Local regulations may restrict where you can drill and how much water you can draw.

- Drilling: The depth of your well will depend on the water table in your region. Some wells may only need to be 50 feet deep, while others could require drilling several hundred feet.

- Pumps and Pressure Tanks: Once your well is drilled, you'll need a pump to bring the water to the surface and a pressure tank to store and distribute it throughout your home.

Natural Springs

If your land has access to a natural spring, you have an excellent water source that often requires little intervention. Springs occur when groundwater rises to the surface, and they can provide a steady flow of fresh water.

How to make use of a spring:

- Spring Capture: Protecting the spring from contamination is crucial. Build a spring box (a concrete or stone structure) to enclose the spring's source and direct the water into a pipe or storage tank.

- Filtration: Even though spring water is usually clean, it's still wise to install a filtration system to remove any potential contaminants.

- Gravity-Fed Systems: If your spring is located on higher ground, you can create a gravity-fed water system that eliminates the need for pumps.

Lakes and Rivers

Lakes and rivers can be excellent sources of surface water, but they come with certain challenges, including filtration needs and fluctuating water levels.

- Water Collection: You'll need to build a simple intake system that draws water from the lake or river into your home. This often involves a pipe with a pump or a gravity-fed system, depending on the elevation.

- Filtration: Surface water from lakes and rivers is more likely to contain contaminants like bacteria, algae, and silt. A multi-stage filtration system, including UV sterilization or chemical treatment, may be necessary to ensure the water is safe to drink.

- Environmental Considerations: Ensure that you follow local environmental regulations when drawing water from lakes or rivers. Over-extraction could damage local ecosystems.

2. Groundwater vs. Surface Water: Choosing the Right Source

When choosing a water source, one of the first decisions you'll need to make is whether to rely on groundwater (wells) or surface water (lakes, rivers, streams). Each option has its pros and cons, so it's important to understand how each works and which will be the most reliable for your situation.

Groundwater (Wells)

Groundwater, accessed through wells, is generally considered a more stable and reliable source of water for long-term off-grid living. Groundwater is less susceptible to seasonal changes and contamination compared to surface water, making it a preferred choice for many.

Advantages of Groundwater:

- Consistency: Groundwater supplies tend to be less affected by drought or seasonal fluctuations. Once you've tapped into a well, you generally have a continuous supply of water, as long as the water table doesn't drop too low.

- Cleaner Water: Groundwater is naturally filtered as it moves through soil and rock, reducing the likelihood of contamination from surface pollutants like chemicals, animal waste, or bacteria.

- Low Maintenance: After the initial drilling and installation, wells require minimal upkeep. Periodic water testing and pump maintenance are usually sufficient.

Challenges of Groundwater:

- Cost of Drilling: Installing a well can be expensive, especially if you need to drill deep to reach the water table. In some cases, a well might cost several thousand dollars.

- Location-Specific: Not all locations have easy access to groundwater. In some areas, the water table may be too deep, or the water quality may be poor, requiring expensive filtration systems.

- Water Rights: In some regions, there are legal restrictions on how much groundwater you can draw. Be sure to check local regulations before drilling a well.

Surface Water (Lakes, Rivers, Streams)

Surface water can be a great resource for those living near lakes, rivers, or streams. However, it requires more infrastructure and treatment to ensure the water is safe for drinking.

- Advantages of Surface Water:

- Accessibility: Surface water is easier to access, especially if you live near a body of water. You won't need to drill or install expensive infrastructure to get started.

- Abundant Supply: In regions with large lakes or rivers, surface water can provide an almost unlimited source of water, as long as it's properly managed.

- Simple Filtration: For uses like irrigation or livestock watering, surface water may require little to no filtration.

- Challenges of Surface Water:

- Contamination Risk: Surface water is much more likely to be contaminated with bacteria, chemicals, or other pollutants. As a result, it requires more extensive filtration and treatment to be safe for drinking.

- Seasonal Variability: Surface water levels can fluctuate dramatically with the seasons, and some rivers or streams may dry up completely during droughts. In colder climates, lakes and rivers can freeze over, cutting off access to water.

- Legal and Environmental Considerations: There may be restrictions on how much water you can draw from a public water source like a river or lake, and overuse could harm local ecosystems.

Ultimately, the decision between groundwater and surface water will depend on your location, the availability of each resource, and your long-term water needs. Many off-grid homesteads use a combination of both sources to ensure a reliable supply.

5. Water Collection Systems

If your property doesn't have easy access to natural water sources, or you want to supplement your supply, water collection systems can provide an effective solution. The most common method is rainwater harvesting, which involves capturing and storing rainwater for later use. Whether you're in an area with heavy rainfall or just occasional showers, rainwater collection can help make your off-grid life more sustainable.

Rainwater Harvesting: How to Set Up a System

Rainwater harvesting is probably among the simplest and most cost-effective methods for collecting water. It's a versatile solution that works in a wide variety of climates, and with the right setup, it can provide all the water you need for drinking, irrigation, and household use.

Step 1: Roofing and Gutter Setup

The roof of your home (or outbuildings) is the first part of any rainwater harvesting system. Rainwater runs off the roof and into gutters, where it is directed into storage tanks. Here's how to set up an efficient system:

- Roof Materials: The material of your roof will affect the quality of the water you collect. Metal or tile roofs are ideal because they are smooth and easy to clean, reducing the risk of contamination. Avoid using asphalt shingles, as they may leach chemicals into the water.

- Gutters and Downspouts: Install gutters along the edges of your roof to collect rainwater as it falls. Make sure the gutters are sloped properly so that water flows smoothly into the downspouts. Attach mesh screens or gutter guards to keep out leaves and debris.

- First-Flush Diverter: Install a first-flush diverter at the downspout to prevent the first few gallons of rainwater, which may contain dirt or contaminants from the roof, from entering your storage system. The diverter will automatically discard this initial runoff and then allow the cleaner water to flow into your storage tanks.

Step 2: Water Storage Options

Once the rainwater has been collected, it needs to be stored in a tank or cistern. The size and type of your storage system will depend on your water needs, the amount of rainfall in your area, and how much space you have available.

-Barrels: For small-scale rainwater harvesting, barrels are an easy and affordable option. They can be placed under your downspouts and hold anywhere from 50 to 100 gallons of water. Barrels are great for gardening and irrigation, but may not provide enough water for drinking and household use.

- Cisterns: For larger-scale storage, cisterns are a better choice. These are often made from plastic, concrete, or metal, and can hold several thousand gallons of water. Cisterns can be installed above ground or buried underground to save space and keep the water cool.

- Modular Tanks: Modular water tanks are another flexible option. These tanks are designed to be stackable or placed side by side, allowing you to expand your storage capacity as needed. They are often made from durable plastic or metal and can be fitted with filtration systems.

Step 3: Filtering and Purifying Rainwater

Even though rainwater is generally clean, it still needs to be filtered and purified before drinking. Contaminants like dirt, bird droppings, or pollutants from the atmosphere can make their way into the water, so it's important to have a multi-stage filtration system.

- Pre-Filtration: Before the water enters your storage tank, it should pass through a simple filter like a mesh screen or sediment trap to remove large debris.

- Sediment Filters: Install a sediment filter to remove small particles like dust or sand from the water. This is usually a cartridge filter that can be cleaned or replaced regularly.

- Activated Carbon Filters: For drinking water, use an activated carbon filter to remove chemicals, pesticides, and odors. This will improve the taste and safety of the water.

- UV Light or Chemical Treatment: Finally, purify the water using UV light or chemical disinfectants like chlorine or iodine. This step will kill any remaining bacteria or viruses, making the water safe to drink.

By implementing these water sourcing and collection systems, you'll ensure a steady supply of clean water for your off-grid home. Whether you're tapping into natural sources like wells and rivers or setting up a rainwater harvesting system, careful planning and installation will help you maintain water independence. With a reliable water system in place, you can focus on the other aspects of self-sufficiency and sustainable living.

6. Water Purification Systems, Making Water Safe for Off-Grid Living

After sourcing water from wells, rivers, or rainwater harvesting systems, ensuring that it's safe for drinking and daily use is critical to maintaining health in an off-grid home. While natural water sources are often reliable, they can contain contaminants such as sediment, bacteria, chemicals, and other harmful microorganisms. Proper purification ensures that the water you consume is free from pathogens and safe for long-term use. This section will cover different filtration methods and disinfection techniques that are practical for off-grid living, giving you clear steps to create clean, potable water.

Filtration Methods: Removing Sediment and Contaminants

Filtration is the first step in any water purification system. It removes larger particles, like dirt and sediment, as well as certain chemicals and biological contaminants, improving both the clarity and safety of your water. Depending on your water source and the level of contamination, different types of filters may be more or less appropriate for your setup.

1. Slow Sand Filters: Natural, Low-Cost Filtration

Slow sand filtration is a simple yet effective method for treating water, relying on the natural biological processes that occur as water percolates through layers of sand. This method is low-cost and sustainable, making it ideal for off-grid systems where energy and resources are limited.

- How It Works: A slow sand filter consists of a container filled with layers of fine sand and gravel. Water passes slowly through the sand, where microorganisms form a biofilm (called the "schmutzdecke") that traps and breaks down harmful bacteria and organic material. The sand further removes fine particles and some chemical contaminants.

- Advantages:

- Low Maintenance: Once set up, a slow sand filter requires very little maintenance. The filter can operate for several years without needing to be replaced, though occasional cleaning of the top sand layer may be necessary.

- No Chemicals or Electricity: Slow sand filters rely on natural biological processes and gravity, making them a fully passive system that doesn't require any external power source or chemicals.

- Durable and Long-Lasting: The materials required (sand and gravel) are inexpensive and widely available, and the system can be built and maintained by hand.

- Limitations:

- Slow Process: As the name suggests, the water filtration process is slow, which means you'll need a large filter and plenty of time for water to pass through it if you're supplying a household.

- Effectiveness on Very Polluted Water: Slow sand filters are not ideal for heavily polluted water, particularly water contaminated with chemicals or high levels of microorganisms. They are best used in conjunction with other purification methods.

Practical Tip: For best results, place your slow sand filter at a higher elevation than your water storage tank so you can create a gravity-fed system, reducing the need for pumps.

2. Activated Carbon Filters: Removing Chemicals and Odors

Activated carbon (or charcoal) filters are highly effective at removing chemical contaminants, unpleasant odors, and tastes from water. These filters work by adsorbing impurities onto the surface of activated carbon, trapping them inside the filter media.

- How It Works: Activated carbon is highly porous, giving it an extensive surface area that binds to organic compounds, chlorine, pesticides, and other harmful chemicals as water passes through the filter. The result is cleaner, better-tasting water, though activated carbon filters do not remove viruses or bacteria.

- Advantages:

- Effective at Removing Chemicals: Activated carbon excels at filtering out chemical contaminants, including chlorine, volatile organic compounds (VOCs), and certain pesticides that other filtration methods might miss.

- Improves Taste and Odor: If your water has a strong chlorine taste, metallic flavor, or musty smell, activated carbon filters will help improve the water's overall quality.

- Relatively Inexpensive: These filters are widely available, affordable, and easy to install in most water systems.

- Limitations:

- Not Effective for Pathogens: Activated carbon does not remove bacteria, viruses, or minerals from water, so it needs to be paired with other filtration or disinfection methods for safe drinking water.

- Regular Replacement Needed: Over time, the carbon's adsorption capacity diminishes, meaning the filter must be replaced regularly to maintain effectiveness.

Practical Tip: Combine activated carbon filtration with a UV or chlorine-based disinfection system for a comprehensive water purification setup.

3. Reverse Osmosis Systems: Advanced Filtration for Maximum Purity

Reverse osmosis (RO) is a powerful filtration method that removes nearly all contaminants from water, including salts, heavy metals, and microorganisms. It's one of the most effective ways to purify water, though it requires a more complex setup and regular maintenance.

- How It Works: In a reverse osmosis system, water is forced through a semipermeable membrane that filters out particles as small as 0.0001 microns. The system removes up to 99% of dissolved salts, bacteria, viruses, and other impurities. The clean water passes through the membrane, while contaminants are flushed out as waste.

- Advantages:

- Comprehensive Filtration: RO systems can remove a wide range of contaminants, including heavy metals (like lead and arsenic), dissolved salts, and harmful pathogens, making it one of the most effective filtration methods available.

- Excellent for Hard Water: If your water is mineral-heavy or has a high salt content, RO is one of the few filtration methods capable of producing truly soft, clean water.

- Improves Taste and Quality: The water produced by an RO system is exceptionally clean and often tastes better than any other filtration method.

- Limitations:

- Water Wastage: RO systems produce wastewater, as not all the water that passes through the membrane is usable. For every gallon of clean water, an RO system may discard two to three gallons as waste.

- Cost and Complexity: These systems can be expensive to install and require electricity to operate. Additionally, the membranes need to be replaced periodically, which adds to maintenance costs.

Practical Tip: If you're concerned about water wastage, consider using the rejected water from your reverse osmosis system for non-potable uses like irrigation or flushing toilets

Disinfection: Eliminating Pathogens from Water

While filtration removes physical impurities from water, it often isn't enough to ensure that water is free of harmful bacteria, viruses, and parasites. Disinfection is the next critical step, killing off any pathogens that might remain after filtration. There are several tried-and-true methods to disinfect water, each with its pros and cons depending on your off-grid setup.

1. Boiling: The Simplest Method

Boiling water is one of the oldest and most reliable methods of disinfection. It's easy to do with basic equipment and is effective at killing most bacteria, viruses, and parasites that may be present in your water.

- How It Works: Heat kills pathogens by denaturing their proteins and disrupting cell membranes. To ensure that your water is safe to drink, bring it to a rolling boil for at least one minute (three minutes if you're above 6,500 feet in altitude, where water boils at a lower temperature).
  - Advantages:
    - Simple and Effective: Boiling is easy to do with nothing more than a heat source, making it a good option for emergency situations or backup disinfection.
    - No Special Equipment Needed: If you have access to fire, gas, or electricity, boiling can be done quickly with materials already at hand.
  - Limitations:
    - Time and Energy-Consuming: Boiling requires fuel and time, which may not be practical if you need to disinfect large amounts of water or have limited energy resources.
    - Does Not Remove Chemicals: Boiling doesn't remove chemical contaminants like heavy metals or pesticides, so it should be used in conjunction with filtration methods.

Practical Tip: Always let boiled water cool in a covered, clean container before drinking to avoid recontamination.

2. Chlorination: Chemical Disinfection for Large-Scale Use

Chlorination is a common and effective method of disinfecting water, especially in larger quantities. It works by introducing a small amount of chlorine (usually in the form of household bleach or chlorine tablets) to kill pathogens.

- How It Works: Chlorine reacts with water to form hypochlorous acid, which disrupts the cell walls of bacteria and viruses, killing them quickly. To use chlorine, add about 8 drops of unscented household bleach per gallon of water, stir well, and let it sit for at least 30 minutes before consuming.
  - Advantages:
    - Inexpensive and Widely Available: Chlorine, either in the form of bleach or chlorine tablets, is easy to obtain and affordable, making it a practical option for off-grid living.
    - Effective Against Most Pathogens: Chlorine kills most bacteria, viruses, and parasites, ensuring your water is safe to drink.
    - Easy to Scale: Chlorination is an ideal method for disinfecting large quantities of water, whether you're treating a household supply or water stored in large cisterns.
  - Limitations:
    - Taste and Odor: Chlorinated water often has a distinct taste and smell, which may be unpleasant, though this can be reduced by using activated carbon filters.

- Requires Precise Dosing: Too little chlorine won't disinfect the water effectively, while too much can be harmful to your health. Follow the recommended dosing instructions carefully.

- Does Not Remove Other

Contaminants: Like boiling, chlorination doesn't remove chemical pollutants or sediment.

Practical Tip: If you're using chlorine, make sure your bleach is unscented and free of additives. Also, rotate your bleach supply regularly, as its effectiveness degrades over time.

3. UV Light: High-Tech Water Sterilization

Ultraviolet (UV) light is a high-tech solution for disinfecting water, using light to kill bacteria, viruses, and other pathogens. It's an excellent option for off-grid systems that have access to solar or battery power.

- How It Works: UV water purifiers emit ultraviolet light at a wavelength that disrupts the DNA of microorganisms, rendering them unable to reproduce and thus effectively killing them. These devices are usually installed in your water line or used as portable handheld units.
- Advantages:
- Highly Effective: UV light is extremely effective at neutralizing a wide range of pathogens, including bacteria, viruses, and parasites.
- No Chemicals or Taste Alteration: UV sterilization doesn't require chemicals, so it won't change the taste, smell, or appearance of your water.
- Low Maintenance: UV systems are easy to maintain, requiring little more than occasional cleaning of the UV bulb or quartz sleeve.
- Limitations:
- Requires Electricity: UV systems need a reliable source of electricity, which can be a challenge in off-grid settings unless you have a dependable solar, wind, or battery system in place.
- Does Not Remove Particulates: Like other disinfection methods, UV light won't remove physical particles, heavy metals, or chemicals, so it must be used in conjunction with filtration systems.

Practical Tip: If you're using a UV sterilizer, ensure that the water is clear and free of sediment before treatment, as cloudy water can block the light and reduce effectiveness.

Choosing the Right System for Your Needs

Each method of filtration and disinfection has its strengths and limitations, and the right choice depends on your specific water source, location, and available resources. Many off-grid homeowners use a combination of methods to ensure their water is both physically filtered and biologically safe. For example, a slow sand filter paired with UV sterilization offers a low-maintenance solution that requires no chemicals or electricity, while reverse osmosis combined with chlorination provides an ultra-clean system that's ideal for purifying even the most contaminated water sources.

By investing in the right water purification systems, you can ensure that your off-grid home has a steady supply of clean, safe drinking water, giving you peace of mind and independence from external utilities.

7. Water Distribution Systems, Efficient Solutions for Moving Water Off-Grid

Once you've secured a reliable and purified water source, the next step is distributing that water to your home, garden, or livestock. Off-grid water distribution systems can vary greatly depending on your specific setup, water needs, and energy resources. This section will cover two main methods of water distribution: gravity-fed systems and pumping systems. Whether you're taking advantage of the natural elevation of your property or using pumps powered by solar or electricity, this guide will help you design a system that's efficient, sustainable, and reliable.

Gravity-Fed Water Systems: Harnessing the Power of Elevation

Gravity-fed water systems are one of the simplest and most energy-efficient ways to distribute water in an off-grid environment. By using the natural force of gravity, these systems can move water from a higher elevation (such as a water storage tank or spring) down to where it's needed, without the need for pumps or electricity. If your property has natural elevation changes or if you can elevate your water storage, a gravity-fed system can be an ideal solution for distributing water to your home, garden, or other areas.

1. How Gravity-Fed Systems Work

Gravity-fed systems rely on the principle that water will naturally flow from a higher elevation to a lower one. By placing your water source or storage tank at a higher elevation than the areas you need to supply (such as your home or garden), gravity will pull the water downhill through pipes or hoses. The steeper the slope, the faster the water will flow.

Basic Components of a Gravity-Fed System:

- Water Source: This could be a rainwater storage tank, a natural spring, or any other source located uphill.

- Piping: PVC or polyethylene pipes are commonly used to transport water from the source to your home or garden. The diameter of the pipes will affect water pressure and flow rate.

- Pressure Regulation: Since gravity-fed systems can generate high pressure depending on the elevation difference, pressure regulators may be needed to prevent pipes from bursting and to control water flow.

- Valves and Taps: These allow you to control where the water flows and how much is released at various points in your system.

2. Designing a Gravity-Fed System

Designing an efficient gravity-fed system requires careful planning to ensure that water is delivered where it's needed with adequate pressure. Let's break down the key steps to designing a successful gravity-fed water distribution system.

Step 1: Identify Your Water Source and Elevation

Start by identifying your water source, such as a storage tank, well, or spring, and note its elevation relative to the areas where you need water. The greater the elevation difference, the stronger the water pressure will be. A minimum elevation difference of 10-15 feet is recommended for a basic gravity-fed system.

Step 2: Calculate Water Pressure

Water pressure in a gravity-fed system is determined by the height difference between the water source and the outlet. For every foot of elevation drop, you gain approximately 0.43 pounds per square inch (psi) of pressure. If your tank is 50 feet above your home, for example, you'll have roughly 21 psi, which is enough to operate faucets and basic irrigation systems.

Step 3: Choose the Right Pipe Size

The size of the pipes you use will affect both the water pressure and the flow rate. Larger pipes reduce friction and allow more water to flow, but they also reduce pressure. For most gravity-fed systems, a 1-inch to 2-inch diameter pipe is sufficient for household use, but larger pipes may be needed for irrigation or livestock watering.

Step 4: Install Pressure Regulators and Taps

If the elevation difference is significant, the water pressure may be too high for safe use in your home. Install pressure regulators at key points in the system to reduce pressure to a manageable level. Taps and valves allow you to control where the water flows, making it easy to direct water to different areas as needed.

3. Advantages of Gravity-Fed Systems

Gravity-fed systems are incredibly efficient and sustainable, making them a popular choice for off-grid water distribution. Here's why they're an excellent option:

- Energy-Free Operation: Once set up, gravity-fed systems don't require any electricity or fuel to operate. Water flows naturally downhill, making it one of the most energy-efficient ways to distribute water.

- Low Maintenance: With no moving parts, gravity-fed systems are less prone to breakdowns and require minimal maintenance. Occasional cleaning of pipes and valves is usually all that's needed to keep the system running smoothly.

- Cost-Effective: Since gravity-fed systems don't require pumps or other mechanical equipment, the initial setup costs are lower, and ongoing costs are minimal.

- Reliability: As long as you have a consistent water source and sufficient elevation, gravity-fed systems are reliable and won't be affected by power outages or fuel shortages.

4. Limitations of Gravity-Fed Systems

While gravity-fed systems are highly efficient, they do have some limitations:

- Dependence on Elevation: The biggest drawback of gravity-fed systems is that they require a significant elevation difference to generate adequate pressure. If your property is flat or your water source is at the same level as your home, a gravity-fed system won't work without elevating your storage tanks.

- Pressure Variations: Water pressure can fluctuate depending on the elevation difference and the distance the water has to travel. Long runs of pipe or sharp bends can reduce pressure and flow rate.

- Limited to Downhill Distribution: Gravity-fed systems only work for moving water downhill. If you need to move water uphill or across flat terrain, you'll need a pump.

Pumping Systems: Powered Solutions for Water Distribution

When gravity isn't an option, pumping systems provide an effective way to move water around your property. Whether you're drawing water from a well or distributing it from a storage tank, pumps can provide the pressure needed to move water uphill or across long distances. Off-grid pumping systems can be powered by manual effort, solar energy, or electricity, depending on your energy setup.

1. Hand Pumps: Simple, Manual Solutions

Hand pumps are a low-tech, manual solution for drawing water from wells or storage tanks. They don't require any electricity, making them a reliable option for off-grid systems, particularly in areas where power is limited or unavailable.

How Hand Pumps Work: Hand pumps operate by manually lifting water to the surface through a series of mechanical motions. They can be installed on wells, cisterns, or water tanks, and are ideal for shallow to moderately deep water sources.

Advantages:

- No Power Required: Hand pumps operate purely on human power, so they're immune to power outages or system failures.

- Low-Cost: These pumps are inexpensive to purchase and install, making them an affordable option for small-scale water distribution.

- Reliable in Emergencies: Hand pumps are a great backup option in case of pump failure or power shortages.

Limitations:

- Labor-Intensive: Hand pumps require physical effort to operate, which may not be practical for large quantities of water or frequent use.

- Limited Depth: Most hand pumps are effective for wells or tanks no deeper than 50-100 feet. For deeper wells, a more powerful pump will be needed.

Practical Tip: Keep a hand pump as a backup system for emergencies or for small-scale water needs, such as watering a garden or providing water to livestock.

2. Solar-Powered Pumps: Sustainable and Automated Water Distribution

Solar-powered pumps are an excellent option for off-grid water distribution, providing a sustainable, energy-efficient solution that uses solar panels to power the pump. These systems can be used to draw water from wells, cisterns, or lakes and can distribute water to your home, irrigation systems, or livestock areas.

How Solar-Powered Pumps Work: Solar panels capture sunlight and convert it into electricity, which powers the water pump. Solar-powered pumps come in various sizes and capacities, making them suitable for different water needs, from small garden irrigation systems to large household water supplies.

Advantages:

- Energy Independence: Solar-powered pumps operate entirely on renewable energy, reducing your reliance on external power sources or fuel.

- Automated Operation: Unlike hand pumps, solar-powered pumps operate automatically, requiring little to no manual effort once installed.

- Scalable: Solar pumps are available in a range of capacities, so you can scale your system based on your water needs. For larger homes or farms, you can install additional solar panels or upgrade to a more powerful pump.

Limitations:

- Weather Dependent: Solar-powered pumps rely on consistent sunlight to operate efficiently. In cloudy or rainy conditions, the pump may not perform at full capacity unless you have a battery storage system in place.

- Upfront Costs: While solar pumps save money on energy costs in the long term, the initial investment in solar panels, batteries, and the pump itself can be high.

- Requires Regular Maintenance: Solar panels and pumps require occasional maintenance, including cleaning the panels and checking electrical connections.

Practical Tip: Pair your solar-powered pump with a battery storage system to ensure you have consistent water distribution, even on cloudy days or at night.

3. Electric Pumps: High-Powered Solutions for Deep Wells and Large-Scale Use

Electric pumps offer the most power and versatility for water distribution, especially when dealing with deep wells or large volumes of water. These pumps are powered by traditional electric grids, generators, or off-grid systems like solar or wind power, and they can handle complex distribution needs.

How Electric Pumps Work: Electric pumps use motors to draw water from deep wells or tanks and distribute it through pipes to your home, irrigation system, or other areas. Submersible electric pumps are often used for deep wells, while surface electric pumps are better suited for shallow wells or storage tanks.

Advantages:

- Powerful and Efficient: Electric pumps provide consistent water pressure and can move large quantities of water quickly, making them ideal for deep wells or high-demand systems.

- Automatic Operation: Most electric pumps can be connected to pressure switches, allowing the system to automatically turn on and off as needed, providing hands-free water distribution.

- Versatile: Electric pumps can be used for a variety of applications, from household water supply to large-scale irrigation or livestock watering.

Limitations:

Dependence on Power: Electric pumps require a reliable power source, which can be a challenge in off-grid settings. Backup systems, like generators or battery storage, are often needed in case of power outages.

- Higher Energy Costs: If you're using a generator or grid electricity to power your pump, energy costs can add up, particularly for large systems.

- Complex Installation: Installing an electric pump, especially a submersible pump for a deep well, can be more complex and expensive compared to other systems.

Practical Tip: If you have a deep well or large water needs, consider using a hybrid system that combines an electric pump with solar power or backup hand pumps for greater flexibility and reliability.

Choosing the Right System for Your Needs

When it comes to distributing water in your off-grid setup, the best system will depend on the specific characteristics of your property, your water source, and your energy availability. Gravity-fed systems are ideal for properties with natural elevation, while hand pumps and solar-powered pumps provide low-energy, sustainable solutions for smaller setups. For deeper wells or larger water demands, electric pumps offer the power and efficiency needed, though they may require additional energy management.

By carefully selecting and designing your water distribution system, you can ensure that your off-grid home remains self-sufficient, with a steady and reliable supply of water to meet all your household, agricultural, and livestock needs.

8. Wastewater Management Systems, Sustainable Solutions for Off-Grid Living

Managing wastewater effectively is a critical component of off-grid living. An efficient wastewater system not only ensures sanitation and environmental protection but also helps conserve water—a valuable resource in any off-grid setup. In this section, we will explore two primary wastewater management solutions: greywater systems, which recycle water from sinks and showers for reuse, and septic systems, which handle household sewage. Whether you plan to build your system from scratch or hire professionals, these systems are essential for maintaining hygiene and sustainability off the grid.

Greywater Systems: Recycling Water for Sustainable Use

Greywater systems are an excellent way to reuse water from everyday household activities such as washing dishes, showering, or laundry. Instead of letting this water go to waste, greywater systems treat and redirect it for other purposes, such as irrigating gardens or flushing toilets. This not only conserves water but also reduces the overall strain on your water supply, which is especially important in off-grid settings where every drop counts.

1. What is Greywater?

Greywater refers to the somewhat clean wastewater that comes from sinks, showers, bathtubs, and washing machines. Unlike blackwater (from toilets), greywater contains fewer harmful pathogens and can be safely reused in certain applications with minimal treatment.

Sources of Greywater:

- Bathroom sinks and showers: These produce greywater that is generally clean, though it may contain soap, shampoo, or dirt.

- Laundry: Washing machines produce greywater that may contain detergents or fabric softeners.

- Kitchen sinks: Some systems avoid using water from kitchen sinks, as it may contain food particles, fats, and oils that complicate filtration.

Why Recycle Greywater?:

- Conserve Fresh Water: Recycling greywater for irrigation or other non-potable uses significantly reduces your demand for clean, drinkable water.

- Reduce Wastewater Output: A well-designed greywater system reduces the amount of water you send to a septic system or sewage treatment plant, which can extend the life of your septic tank and reduce maintenance needs.

- Improve Sustainability: Greywater systems promote sustainable water use, particularly in areas prone to drought or where water resources are scarce.

2. How Greywater Systems Work

Greywater systems can vary from simple setups that reroute water from a washing machine to more complex systems that treat greywater and distribute it through underground irrigation lines. Let's explore how these systems are set up and operated.

Basic Components of a Greywater System:

- Collection: Greywater is first collected from sinks, showers, or washing machines. This can be as simple as installing diverters to redirect the water into collection tanks or pipes.

- Filtration: Simple filtration methods, such as mesh screens, are used to remove large particles like hair, dirt, or food scraps.

- Distribution: Once filtered, the greywater can be distributed directly to where it's needed. Common uses include irrigation for gardens, orchards, or landscaping.

- Storage Tanks: In more advanced systems, greywater is stored in tanks until it is needed. However, greywater should not be stored for long periods as it can become a breeding ground for bacteria and odors.

Designing a Greywater System:

Step 1: Identify Water Sources and Needs

Begin by evaluating your greywater sources and estimating how much water your household produces. On average, a family of four generates about 100 to 150 gallons of greywater per day from bathing and washing alone. Identify how much of this water you can realistically reuse for purposes like irrigation.

Step 2: Choose a Greywater Use

Decide how you plan to reuse the greywater. The most common use is for landscape irrigation. Greywater can be used to water plants, trees, and shrubs, as long as it's delivered below the soil surface. Avoid using greywater on edible plants like vegetables unless you have a more advanced filtration and disinfection system.

Step 3: Install Filtration and Distribution Components

Set up a filtration system that suits your needs. For small systems, a simple mesh filter may suffice. More complex systems might include sediment filters or biological filters that further treat the greywater. Install pipes or hoses to direct the water to your irrigation areas. Make sure the water is distributed below the soil surface to avoid contact with humans and animals.

3. Types of Greywater Systems

There are several types of greywater systems, each with varying levels of complexity and treatment. Here are three common options for off-grid living:

1. Simple Diversion Systems

Simple diversion systems are the most basic form of greywater recycling. These systems use diverters or valves to direct greywater directly from the source (such as a sink or washing machine) to a nearby garden or landscape.

- How It Works: The system collects greywater directly from the drainpipe and redirects it to irrigation lines or a nearby garden bed. No filtration or treatment is required, making it a cost-effective and easy-to-install option.

- Best For: Low-maintenance setups where greywater will be used immediately for irrigation, such as in a garden or orchard.

- Limitations: These systems do not store water, meaning the greywater must be used as it's produced. Additionally, they may not be suitable for use in areas where you plan to grow food.

2. Branched Drain Systems

Branched drain systems are a more advanced setup that distributes greywater to multiple areas using a series of pipes and valves. These systems are ideal for large properties where water needs to be spread across various zones, such as gardens, orchards, or flower beds.

- How It Works: Water is collected from multiple sources (sinks, showers, etc.) and directed through a network of pipes that branch out to different irrigation zones. The water is typically delivered through small trenches or mulch basins that allow it to percolate into the soil.

- Best For: Off-grid properties with large gardens, trees, or orchards where irrigation needs to be spread out over a large area.

- Limitations: Branched drain systems can be more complex to install, requiring careful planning to ensure even distribution. Regular maintenance is needed to keep the pipes clear of debris.

3. Constructed Wetlands

Constructed wetlands are an eco-friendly greywater treatment system that uses plants and natural filtration to clean and recycle water. This system mimics natural wetlands, where plants and microorganisms filter and purify the water.

- How It Works: Greywater flows into a shallow, gravel-filled basin planted with wetland vegetation like reeds, cattails, or other water-loving plants. The plants and gravel filter out contaminants while the roots absorb nutrients, cleaning the water as it flows through the system. The treated water can then be used for irrigation or other non-potable uses.

- Best For: Off-grid homes looking for a natural, low-maintenance solution to greywater recycling, particularly in areas where aesthetics and environmental impact are important.

- Limitations: Constructed wetlands require space and proper design to ensure effective filtration. They also need occasional maintenance to keep plants healthy and remove any build-up of debris or sediment.

Practical Tip: When installing a greywater system, avoid using harsh chemicals or cleaners in your sinks, showers, or laundry. Opt for biodegradable soaps and detergents to protect the environment and ensure your greywater is safe for reuse.

Septic Systems: Handling Household Sewage Safely

While greywater systems handle water from sinks and showers, septic systems manage blackwater — wastewater that contains human waste from toilets. For off-grid living, a septic system is essential for managing household sewage in a sanitary and environmentally safe way. Septic systems come in different types, and you can choose between building a DIY system or hiring professionals for installation. Regular maintenance is crucial to ensure the system continues to function properly over time.

1. How Septic Systems Work

A septic system is an underground wastewater treatment structure that uses natural processes to treat and dispose of household sewage. The system consists of two primary components: a septic tank and a drain field.

- Septic Tank: The septic tank is a watertight container buried underground. Wastewater flows from the house into the tank, where solids settle to the bottom, forming sludge, while oils and grease float to the top, forming scum. Bacteria in the tank break down the waste, reducing its volume.

- Drain Field: The liquid wastewater (called effluent) flows from the tank into a drain field or leach field. The effluent is dispersed into the soil, where it is further treated by natural biological processes before reentering the groundwater system.

2. DIY vs. Professional Septic Systems

When it comes to installing a septic system, you have the option of building a DIY system or hiring professionals to install a more complex system. Let's explore the pros and cons of each.

DIY Septic Systems

For those with some construction skills, building a DIY septic system can be a cost-effective solution. DIY systems are often simpler and designed for smaller households or remote cabins.

- How It Works: A basic DIY septic system consists of a homemade septic tank (often made from plastic or concrete) connected to perforated pipes that disperse the effluent into a leach field. Gravel or sand is used to promote drainage and filter the effluent as it moves through the soil.

Advantages:

- Lower Cost: DIY systems are significantly cheaper to install than professional systems, making them an appealing option for small, off-grid homes.
- Simplicity: These systems are generally smaller and less complex, making them easier to maintain for those familiar with basic plumbing and construction.

Limitations:

- Regulatory Challenges: Many areas have strict regulations regarding the installation of septic systems. DIY systems may not meet local codes, so it's important to check with your local authorities before building one.
- Capacity: DIY septic systems are typically designed for smaller households and may not be sufficient for larger families or properties with heavy water usage.
- Maintenance: DIY systems require more hands-on maintenance to ensure that they don't clog or fail.

Professional Septic Systems

Professional septic systems are installed by licensed contractors and are designed to handle larger volumes of wastewater. These systems meet all local building codes and are more suitable for permanent homes or larger properties.

- How It Works: A professional system includes a precast concrete or fiberglass septic tank and an engineered leach field designed to handle the specific wastewater load of your home. Some systems also include advanced treatment units, such as aeration systems or biofilters, to further treat the effluent.

- Advantages:
- Meets Legal Requirements: Professionally installed systems are designed to comply with local regulations and environmental standards, reducing the risk of fines or penalties.

- Larger Capacity: These systems are built to handle the wastewater needs of larger homes or properties with multiple occupants.

- Low Maintenance: Once installed, professional systems require less hands-on maintenance compared to DIY systems. Regular inspections and pump-outs are typically all that's needed to keep the system functioning properly.

- Limitations:

- Higher Cost: Professional septic systems are more expensive to install, often costing several thousand dollars depending on the size of the system and the complexity of the installation.

- Less Flexibility: Professional systems must be installed according to local codes, which may limit where and how you can build the system on your property.

3. Maintaining a Septic System

Proper maintenance is essential to ensuring that your septic system functions properly for years to come. Neglecting your system can lead to costly repairs, environmental contamination, and health risks.

- Regular Inspections: Have your system inspected by a professional every three to five years to check for leaks, cracks, or other issues.

- Pump-Outs: Septic tanks need to be pumped out regularly to remove the sludge and scum that accumulate over time. How often your tank needs to be pumped depends on the size of the tank and the number of people using it, but most tanks need to be pumped every three to five years.

- Watch What You Flush: Only human waste and toilet paper should be flushed down the toilet. Avoid flushing anything else, including wipes, feminine hygiene products, or grease, as these can clog the system.

- Water Conservation: Reducing water usage helps prevent your septic system from becoming overwhelmed. Spread out laundry loads, fix leaks, and install low-flow fixtures to conserve water and reduce strain on your system.

4. Alternatives to Traditional Septic Systems

For those looking for alternatives to traditional septic systems, there are several eco-friendly options that can be more sustainable for off-grid living:

- Composting Toilets: Composting toilets are a waterless option that converts human waste into compost. These systems are simple to maintain and ideal for remote areas where water conservation is a priority.

- Aerobic Treatment Units (ATUs): These systems use oxygen to break down organic matter more quickly than traditional septic systems. ATUs are often used in areas with poor soil or high water tables, where traditional leach fields are impractical.

- Constructed Wetlands: Similar to greywater wetlands, these systems use plants and microorganisms to treat wastewater before it's dispersed into the environment. Constructed wetlands are an eco-friendly option for homes with limited space for traditional drain fields.

Choosing the Right Wastewater System for Your Off-Grid Home

Whether you're looking to recycle water with a greywater system or handle household sewage with a septic system, it's essential to choose a solution that meets the unique needs of your off-grid lifestyle. Greywater systems offer a sustainable way to conserve water, while septic systems ensure sanitation and safety. By selecting the right combination of systems and maintaining them properly, you can create a sustainable and efficient wastewater management solution for your off-grid home.

Module C | Energy & Power Independence

9. Energy Production, Powering Your Off-Grid Home Efficiently

One of the biggest challenges of off-grid living is ensuring a reliable, continuous energy supply. To maintain self-sufficiency, you'll need to generate, store, and manage energy in ways that are both sustainable and practical for your specific environment. In this section, we will explore various energy production methods suited for off-grid systems, including solar power, wind turbines, hydroelectric power, biomass energy, and hybrid setups. We'll also discuss the importance of backup generators and how to maintain them to ensure your home stays powered even during low energy production periods.

Solar Power Systems: Harnessing the Sun's Energy

Solar energy is one of the most accessible and reliable sources of renewable power for off-grid homes. By installing solar panels, you can generate clean, sustainable energy directly from sunlight, reducing your reliance on traditional fuels or grid electricity.

1. How Solar Power Works

At the heart of any solar power system are photovoltaic (PV) cells, which convert sunlight into electricity. These cells are typically made from silicon and are arranged in solar panels. When sunlight strikes the solar cells, it energizes electrons, generating an electric current. This current, known as direct current (DC), is then transformed into alternating current (AC) by an inverter, allowing it to power household appliances.

2. Installing Solar Panels

Installing solar panels involves careful planning to ensure you capture the maximum amount of sunlight while keeping your system efficient and cost-effective. Here's how to set up a basic solar panel system:

Step 1: Assess Your Energy Needs

Before installing solar panels, calculate your energy consumption. This involves adding up the wattage of all your household appliances and devices and determining how many kilowatt-hours (kWh) you use daily. Energy-efficient appliances will reduce the number of panels required, saving costs.

Practical Tip: Use an energy consumption calculator or track your energy usage for a week to get accurate data on your needs.

Step 2: Determine Your Solar Potential

Solar panels perform best in locations with direct sunlight. To maximize solar gain:

- Roof or Ground Placement: Position your panels on a south-facing roof (in the Northern Hemisphere) or install ground-mounted panels in a sunny area.

- Shading: Avoid installing panels near trees or other obstructions that could cast shadows and reduce solar output.

- Solar Angle: Angle the panels based on your latitude to capture the most sunlight year-round. Adjustable mounting systems can help optimize seasonal performance.

Step 3: Installing the Panels

Once you've chosen the location and size of your solar array, install the panels securely using appropriate mounting hardware. If you're unfamiliar with electrical systems, it's recommended to hire a professional to handle the wiring and inverter installation.

3. Maintaining Solar Panels

Solar panels require very little maintenance, but regular cleaning and inspections will ensure they perform optimally:

- Cleaning: Dust, dirt, and bird droppings can accumulate on your panels, reducing efficiency. Clean them with water and a soft brush every few months, or more often in dusty environments.

- Monitoring Performance: Use a solar monitoring system to track your energy output. If you notice a significant drop in production, it could indicate shading, dirt buildup, or a panel issue.

- Inverter Maintenance: Inverters typically last 10-15 years and may need replacing over the lifespan of your solar system.

Wind Power: Generating Energy with Wind Turbines

Wind power is another renewable energy source that can complement or replace solar energy, especially in areas with strong, consistent winds. Wind turbines convert the kinetic energy of the wind into mechanical energy, which is then used to generate electricity.

1. How Wind Turbines Work

A wind turbine consists of blades, a rotor, a gearbox, and a generator. When the wind blows, it turns the blades, which spin the rotor. This motion is transferred through a gearbox to increase the rotational speed, powering a generator that produces electricity.

2. Installing Wind Turbines

Installing a wind turbine requires careful consideration of the location and wind patterns. Follow these steps for a successful installation:

Step 1: Assess Wind Conditions

Wind turbines are most effective in areas with average wind speeds of 10-12 miles per hour or higher. Use wind mapping tools or conduct a wind survey to measure the wind potential at different heights on your property.

Practical Tip: Install an anemometer to monitor wind speed over several months to determine the best turbine placement.

Step 2: Choose the Right Turbine

Wind turbines come in various sizes, from small residential turbines to large industrial models. For most off-grid homes, a small or medium-sized turbine with a capacity of 1-5 kW is sufficient. Larger turbines can produce more power, but they require more space and higher wind speeds to operate efficiently.

Step 3: Install the Turbine

Wind turbines need to be mounted on tall towers, typically between 30 and 100 feet, to capture the strongest winds. Ensure the tower is tall enough to avoid obstacles like trees or buildings that can cause turbulence. The installation process involves securing the turbine and tower, connecting the generator to your home's electrical system, and installing charge controllers and inverters.

3. Wind Turbine Maintenance

Ongoing maintenance is essential to keep your wind turbine functioning efficiently:

- Inspect the Blades: Check the blades for cracks, debris buildup, or signs of wear. Clean them as needed to prevent efficiency loss.

- Lubricate Moving Parts: Turbines have several moving components, such as the rotor and gearbox, which need regular lubrication to reduce friction and prevent wear.

- Monitor Energy Production: Keep track of your wind energy output. If production drops significantly, inspect the turbine for mechanical issues or changes in wind patterns.

Hydroelectric Power: Small-Scale Water Energy Systems

If you live near a flowing water source like a river or stream, you can generate hydroelectric power using a small-scale turbine system. Hydroelectric power provides a consistent energy source as long as the water flow remains steady, making it ideal for off-grid locations with reliable water access.

1. How Small-Scale Hydroelectric Power Works

Hydroelectric systems use water flow to spin a turbine, which generates electricity in much the same way that wind turbines use air currents. The energy produced is proportional to the flow rate and the "head" (the vertical distance the water falls).

2. Installing a Hydroelectric System

To set up a hydroelectric system, you'll need access to flowing water, the right equipment, and a solid understanding of your power needs.

Step 1: Evaluate Water Flow and Head

Assess the volume of water flowing through the stream or river on your property and the vertical drop (head) that the water will pass through. The greater the flow and head, the more power your system can generate. You can use simple formulas or online calculators to estimate potential energy output.

Step 2: Select a Turbine

There are two main types of turbines for small-scale hydroelectric systems:

- Pelton Wheels: Ideal for high-head, low-flow systems, where the water is channeled to create a high-speed jet that strikes the turbine blades.

- Propeller or Kaplan Turbines: Better for low-head, high-flow systems, where the water flow is steady and slower but still generates sufficient energy.

Step 3: Install the System

Hydroelectric systems require a diversion pipe (also known as a penstock) to channel water from the source to the turbine. The turbine is then connected to a generator, which produces electricity. Depending on the setup, you may also need inverters, batteries, or direct connections to your electrical system.

3. Maintaining a Hydroelectric System

Hydroelectric systems require regular maintenance, especially in regions where seasonal changes can affect water flow:

- Clean the Intake: Debris such as leaves or branches can clog the intake or turbine, reducing efficiency. Regularly clear any blockages.

- Inspect the Turbine: Check the turbine for signs of wear or corrosion, particularly in areas where sediment may cause abrasion.

- Monitor Water Flow: Ensure that water flow remains consistent. If the stream dries up or the flow rate drops significantly, your power generation may be impacted.

Biomass Energy: Using Organic Materials for Power

Biomass energy involves burning or converting organic materials—such as wood, crop waste, or animal manure—into heat or electricity. This method of energy production is sustainable as long as you have access to renewable biomass resources and can manage the environmental impact of burning organic matter.

1. Types of Biomass Energy Systems

There are several ways to produce energy from biomass, including:

- Wood Stoves and Boilers: Wood-burning stoves and boilers are common in off-grid homes. They generate heat for space heating, water heating, and even cooking. Some boilers can also be connected to radiant floor heating systems.

- Biogas Systems: Biogas systems use anaerobic digestion to break down organic matter, such as animal manure or food scraps, into methane. The methane can then be burned to produce heat or electricity.

- Pellet Stoves: Pellet stoves burn compressed wood or biomass pellets, offering a more efficient and cleaner burn than traditional wood stoves.

2. Installing a Biomass Energy System

To set up a biomass system, you'll need to choose the right technology for your needs and have a steady supply of organic material.

Step 1: Choose Your Biomass Source

Determine what types of organic materials are readily available on your property or nearby. If you have access to wood or agricultural waste, a wood stove or biomass boiler may be the best option. If you have livestock or ample food waste, a biogas system could provide an alternative source of energy.

Step 2: Install the System

Wood stoves and boilers require a safe, well-ventilated installation space. Install a chimney or exhaust system to vent smoke and gases outside. Biogas systems need a digester to store organic material, where it will break down and produce methane. Connect the biogas output to a gas burner or generator.

3. Maintaining a Biomass System

Biomass systems require regular maintenance to ensure efficient operation and safety:

- Clean the Stove or Boiler: Regularly remove ash buildup from wood stoves and boilers to maintain efficiency.

- Check the Exhaust: Ensure that the chimney or exhaust system is clear of blockages to prevent smoke or gas buildup.

- Monitor Fuel Supply: Keep an eye on your biomass resources to ensure you have a steady supply of fuel for continuous energy production.

Hybrid Energy Production: Combining Solar, Wind, and Other Sources

For many off-grid homes, relying on a single energy source can be risky, especially in areas where sunlight or wind may be inconsistent. A hybrid energy system combines multiple renewable energy sources—such as solar, wind, and hydro—to create a more reliable and balanced energy supply.

1. Why Use a Hybrid System?

The main advantage of a hybrid system is its flexibility. For example, solar panels produce less energy during cloudy days, but a wind turbine may perform better in those conditions. By combining multiple sources, you reduce the risk of energy shortages.

2. Designing a Hybrid System

Designing a hybrid energy system requires careful planning to balance the strengths and limitations of each energy source. Follow these steps:

Step 1: Assess Local Resources

Determine the renewable resources available in your area. If you have access to abundant sunlight and strong winds, a combination of solar and wind power may be ideal. If you live near a stream or river, hydroelectric power can complement solar or wind energy.

Step 2: Integrate Multiple Systems

Once you've identified your energy sources, integrate them into a single system. This may involve connecting your solar panels, wind turbines, and hydroelectric generator to a central battery bank or inverter. Hybrid inverters are specifically designed to manage multiple power sources and distribute energy efficiently.

3. Maintaining a Hybrid System

A hybrid system requires regular maintenance for each energy source:

- Solar Panels: Clean and inspect regularly.

- Wind Turbines: Lubricate moving parts and check for wear.

- Hydroelectric: Clear debris from the intake and inspect the turbine.

Backup Generators: Ensuring Reliable Power When Needed

Even with a robust renewable energy system, there may be times when energy production is insufficient due to weather conditions, equipment failure, or maintenance. A backup generator provides a reliable source of power during these periods.

1. Types of Backup Generators

There are several types of backup generators available for off-grid homes:

- Diesel Generators: These are powerful and durable, making them suitable for high-demand periods, but they require a steady supply of diesel fuel.

- Propane Generators: Propane is a cleaner-burning fuel than diesel, and propane generators are generally more efficient and easier to maintain.

- Natural Gas Generators: If you have access to a natural gas supply, these generators provide a convenient backup option.

2. Installing and Using a Generator

Backup generators are typically installed outside the home in a well-ventilated area. They should be connected to your home's electrical system via a transfer switch, which automatically switches power from your renewable system to the generator during outages.

3. Maintaining a Generator

- Check Fuel Levels: Keep your fuel tank full, especially during winter or storm seasons when power outages are more likely.

- Regular Testing: Run the generator once a month to ensure it's functioning properly.

- Change Oil and Filters: Regularly change the oil and filters to prevent wear and extend the life of the generator.

Conclusion: Building a Resilient Energy System

By integrating various renewable energy sources—solar, wind, hydroelectric, biomass—into your off-grid home, you can create a resilient and sustainable energy system that meets all your power needs. Adding a backup generator ensures that even during periods of low energy production, your home remains functional. Through proper planning, installation, and maintenance, you can achieve energy independence and live comfortably off the grid.

10. Energy Storage, Powering Your Off-Grid Home When Production is Low

Energy storage is essential for any off-grid living setup. It ensures that power generated by solar panels, wind turbines, or other sources can be used when production is low, such as during cloudy days, calm winds, or nighttime. The two primary forms of energy storage for off-grid systems are battery banks for electricity and thermal storage for managing heat energy. This section will explore how to select, size, and maintain battery banks as well as how thermal storage can be used to store heat for later use.

Battery Banks: Storing Electrical Power Efficiently

Battery banks are the backbone of an off-grid energy system, allowing you to store excess electricity generated during peak production periods and use it when needed. Whether your energy comes from solar, wind, or hydroelectric power, a properly designed battery bank ensures you have a reliable power supply, even when generation stops. Let's dive into the key considerations for selecting, sizing, and maintaining a battery bank.

1. Types of Batteries for Off-Grid Systems

Not all batteries are suitable for off-grid energy storage. Here's an overview of the most common types used for battery banks:

Lead-Acid Batteries

Lead-acid batteries are the most traditional and widely used type of battery for off-grid systems. They are known for their reliability and affordability, though they require maintenance.

- Flooded Lead-Acid (FLA): These batteries require regular maintenance, including checking electrolyte levels and ensuring proper ventilation to prevent gas buildup. They are cheaper but have a shorter lifespan than other types.

- Sealed Lead-Acid (SLA): These batteries are maintenance-free and safer than FLA batteries, as they are sealed and do not emit gas. However, they tend to be more expensive.

Lithium-Ion Batteries

Lithium-ion batteries are gaining popularity thanks to their high energy density, extended lifespan, and low maintenance needs. They are more expensive upfront but provide better performance over time.

Advantages:

- Higher Energy Density: Lithium-ion batteries can store more energy in a smaller space, making them ideal for homes with limited space.

- Longer Lifespan: Lithium batteries typically last 10-15 years, far longer than lead-acid options.

- Minimal Maintenance: Unlike lead-acid batteries, lithium-ion batteries do not require regular maintenance, making them a more convenient choice for off-grid systems.

Nickel-Iron Batteries

Though less common, nickel-iron batteries are extremely durable and can last up to 30 years. They are more expensive and less efficient than lithium-ion or lead-acid batteries but are ideal for those who want a long-lasting, low-maintenance option.

2. Sizing a Battery Bank

Properly sizing your battery bank is crucial to ensuring that you have enough stored energy to meet your household needs. Here's how to calculate the right size for your off-grid system:

Step 1: Determine Your Daily Energy Consumption

Start by calculating your daily energy usage in kilowatt-hours (kWh). This involves adding up the wattage of all the appliances and devices you use and estimating how long each runs per day. For example, if you have a 100-watt light bulb that runs for 10 hours a day, it consumes 1 kWh per day.

Practical Tip: Using energy-efficient appliances and reducing unnecessary energy consumption can significantly reduce the size and cost of your battery bank.

Step 2: Calculate Battery Capacity

Once you know your daily energy consumption, calculate how much storage you'll need by multiplying your daily usage by the number of days you want your battery bank to last without recharging. Most off-grid homes aim for 2-3 days of backup energy.

Example Calculation:

- Daily energy consumption: 10 kWh

- Backup duration: 3 days

- Battery capacity needed: 10 kWh x 3 days = 30 kWh

Step 3: Consider Depth of Discharge

Batteries should not be fully discharged, as this reduces their lifespan. Each type of battery has a recommended depth of discharge (DoD)—the percentage of the battery's capacity that can be used before recharging. Lead-acid batteries typically have a DoD of 50%, while lithium-ion batteries can safely be discharged up to 80-90%.

Adjusted Battery Capacity:

- For a lead-acid battery with a 50% DoD, you would need twice the storage capacity calculated earlier.

- For a lithium-ion battery with an 80% DoD, the required capacity will be slightly larger than the calculated value.

3. Maintaining a Battery Bank

Battery maintenance is key to ensuring the longevity and performance of your storage system. Here's how to maintain different types of batteries:

Lead-Acid Battery Maintenance

- Monitor Electrolyte Levels: For flooded lead-acid batteries, check electrolyte levels every month and top up with distilled water as needed.

- Equalization Charging: Perform equalization charging (a controlled overcharge) every few months to prevent sulfation, which can reduce battery capacity.

- Keep Batteries Clean and Ventilated: Clean the terminals to prevent corrosion and ensure the battery bank is well-ventilated to prevent overheating.

Lithium-Ion Battery Maintenance

- Temperature Control: Lithium-ion batteries perform best within a temperature range of 50-86°F (10-30°C). Install the battery bank in a temperature-controlled environment to extend its lifespan.

- Monitor Charge Cycles: Keep track of the number of charge cycles, as lithium-ion batteries have a finite number of cycles before performance declines.

Nickel-Iron Battery Maintenance

- Low Maintenance: Nickel-iron batteries are extremely durable and require minimal maintenance, though they may need occasional electrolyte replenishment and cleaning.

Thermal Storage: Managing Heat Energy for Heating and Cooling

In addition to electrical energy storage, thermal storage plays a crucial role in off-grid systems, particularly for heating and cooling. Thermal storage systems capture and store heat during times of surplus (such as sunny days) and release it when needed, helping to balance energy use across different seasons or times of day.

1. How Thermal Storage Works

Thermal storage systems work by absorbing and retaining heat energy, which can then be used to heat your home or water when direct energy from your solar panels or biomass system is unavailable. There are several methods for storing thermal energy, each suited for different climates and energy needs.

2. Types of Thermal Storage Systems

1. Water-Based Thermal Storage

Water is an excellent medium for storing thermal energy due to its high heat capacity. Systems that use water for thermal storage can include solar water heaters or radiant floor heating setups.

- **Solar Water Heaters**: These systems use solar panels to heat water, which is stored in insulated tanks for later use. Hot water can be used for domestic purposes or as part of a radiant heating system that distributes heat through pipes embedded in floors or walls.
    - Best For: Homes in sunny climates where hot water is needed year-round.
    - Advantages: Solar water heaters reduce the need for electrical or fuel-based water heating, saving energy and money.
    - Limitations: Solar water heating systems rely on sunlight, so they may not perform well in cloudy or cold climates without additional backup heating methods.

- **Radiant Floor Heating**: In a radiant floor heating system, water is heated and circulated through pipes embedded in the floor. The thermal energy stored in the water is released slowly, providing consistent, even heating throughout your home.
    - Best For: Homes in cold climates where consistent heating is required during the winter.
    - Advantages: Radiant floor heating is highly efficient and provides comfortable, even heat without the need for forced-air systems.
    - Limitations: Installation can be costly and complex, particularly for retrofitting existing homes.

2. Phase Change Materials (PCM)

Phase change materials are innovative thermal storage solutions that absorb and release energy as they change from one phase to another—typically from solid to liquid. These materials are used in specialized thermal storage units that can be integrated into off-grid systems.

- How It Works: When PCM absorbs heat, it melts, storing energy in the process. When the surrounding temperature drops, the PCM solidifies, releasing the stored heat. This process can be repeated indefinitely, making PCM an efficient way to store thermal energy.

- Best For: Homes with variable heating and cooling needs, especially in areas with extreme temperature fluctuations.

- Advantages: PCM systems are compact and can store large amounts of energy in a small space. They also maintain stable temperatures over long periods.

- Limitations: PCM systems are more expensive than traditional thermal storage systems and require professional installation.

3. Thermal Mass Systems

Thermal mass refers to the ability of materials like stone, concrete, or brick to absorb and store heat during the day and release it slowly over time. Buildings designed with thermal mass principles in mind can store solar heat during the day and use it to maintain comfortable indoor temperatures at night.

- How It Works: In a thermal mass system, materials such as concrete floors, brick walls, or stone surfaces are exposed to sunlight during the day. As these materials heat up, they store thermal energy. At night, the heat is released, reducing the need for additional heating sources.

- Best For: Homes in sunny, temperate climates where passive heating and cooling are viable.

- Advantages: Thermal mass systems reduce energy costs by using passive solar heating and cooling techniques. They also improve indoor air quality by reducing the need for forced-air heating systems.

- Limitations: Homes must be designed with thermal mass principles in mind, making retrofits more difficult and costly.

3. Using Thermal Storage for Cooling

While thermal storage is often associated with heating, it can also be used to store cool energy for air conditioning or refrigeration. Some thermal storage systems use ice or other cold-storing materials to capture and store cool energy during the night when temperatures are lower, releasing it during the day for cooling purposes.

Ice Storage Systems

Ice storage systems are used to cool water or air in homes or commercial buildings. These systems work by freezing water at night when electricity is cheaper or more available, and then using the stored ice to cool air or water during the day.

- How It Works: Water is frozen in a large storage tank overnight using energy from your solar or wind system. During the day, the ice melts, cooling air that is circulated through your home or cooling water for refrigeration systems.

- Best For: Homes in hot climates where air conditioning is essential, but electricity is limited.

- Advantages: Ice storage systems reduce the need for continuous electricity usage for air conditioning during the day, making them ideal for off-grid systems with solar panels.

- Limitations: Ice storage systems require careful planning and integration into your overall energy system, and they may not be as efficient in areas where nighttime temperatures remain high.

Selecting the Right Energy Storage System for Your Off-Grid Home

The best energy storage system for your off-grid home depends on several factors, including your climate, energy consumption patterns, and the types of energy you produce. Here's a summary to help guide your decision:

- Battery Banks: Ideal for storing electrical energy from solar, wind, or hydroelectric power systems. Lead-acid batteries are affordable but demand regular maintenance, whereas lithium-ion batteries provide greater efficiency and longer lifespans with minimal upkeep.

- Water-Based Thermal Storage: Best for homes in sunny or cold climates that need to store heat for space heating or hot water. Solar water heaters and radiant floor systems are effective for capturing and using heat energy.

- Phase Change Materials: Suitable for homes with variable energy needs and limited space for storage. PCM systems are highly efficient and can maintain stable indoor temperatures over time.

- Thermal Mass Systems: Perfect for homes designed with passive solar principles, where large surfaces can be used to store and release heat naturally.

Maintaining Your Energy Storage System

Proper maintenance of your energy storage system is essential for ensuring long-term performance and efficiency. Regularly monitor battery charge levels, check for signs of wear or corrosion, and perform any necessary cleaning or fluid top-ups. For thermal storage systems, ensure that your insulation remains intact and check for leaks or blockages in pipes and storage tanks.

Conclusion: Achieving Energy Independence with Effective Storage Solutions

Effective energy storage is the key to a successful off-grid living experience. By choosing the right combination of battery banks and thermal storage systems, you can ensure that your home remains powered, heated, and cooled throughout the year, regardless of energy production fluctuations. Whether you're using a simple battery bank or an advanced PCM system, proper planning and maintenance will help you achieve true energy independence, giving you the freedom to live off the grid without sacrificing comfort or convenience.

11. Energy Distribution, Safely and Efficiently Powering Your Off-Grid Home

Once you have reliable energy production and storage in place, the next step is distributing that energy efficiently throughout your home. Proper energy distribution ensures that all your appliances and systems receive the power they need, while minimizing waste and maximizing safety. In this section, we'll cover the essentials of wiring an off-grid home, understanding voltage considerations, and setting up energy management systems that help you monitor and control your power usage for optimal efficiency.

Wiring Your Home: Distributing Power Safely and Efficiently

Wiring an off-grid home requires careful planning to ensure that the energy generated and stored is distributed effectively throughout your living space. The wiring system must be designed to handle the specific power sources you're using, such as solar panels, wind turbines, or generators, and deliver energy where and when it's needed.

1. Key Considerations for Off-Grid Wiring

Wiring an off-grid home is more complex than wiring a typical grid-connected house due to the variety of power sources, voltage levels, and energy storage systems involved. Here are the key factors to consider:

- Power Source Integration: Off-grid homes often rely on multiple power sources, such as solar panels, wind turbines, and backup generators. Wiring must account for these sources and ensure seamless integration with your energy storage system.

- System Sizing: Your wiring needs to be sized based on the total load your home will place on the system. This includes all appliances, lighting, and other electrical devices, as well as future expansion plans.

- Energy Efficiency: Efficient wiring minimizes energy loss as power is transmitted from your battery bank to various parts of your home. Thicker wires reduce resistance and prevent voltage drops, ensuring that more energy reaches your devices.

Practical Tip: Hire a licensed electrician who is experienced in off-grid systems to help with wiring. Even if you're building your home DIY-style, professional input can prevent costly mistakes and ensure safety.

2. Step-by-Step Guide to Wiring an Off-Grid Home

Follow these steps to set up a safe and efficient wiring system for your off-grid home:

Step 1: Design the Electrical Layout

Start by designing the layout of your electrical system. This includes mapping out where all electrical outlets, switches, and appliances will be located, as well as planning for any future expansion.

- Centralized or Decentralized Distribution: Decide whether you'll use a centralized distribution system (with all wiring converging at a single breaker panel) or a decentralized system, where smaller breaker panels are placed in different parts of the house.

- Wiring for Renewable Energy Systems: If you're using solar, wind, or hydro power, be sure to route the wiring from these sources to your battery bank and inverter first, before distributing power to your home's electrical system.

Step 2: Choose the Right Wiring and Components

Selecting the appropriate wire size, type, and components is essential for a safe and efficient system.

- Wire Gauge: Use thicker wire (lower gauge number) for longer runs to minimize voltage drop. For example, 12-gauge wire is typically used for standard outlets, while thicker 10-gauge or 8-gauge wire may be needed for appliances that draw more power.

- Conduit and Insulation: Use conduit to protect wires from physical damage and ensure that all wires are properly insulated to prevent electrical fires or short circuits.

- Circuit Breakers and Fuses: Install circuit breakers or fuses to protect your wiring and appliances from overloads. Each major appliance or system should have its own dedicated breaker.

Step 3: Grounding the System

Grounding is critical in any electrical system, but especially in off-grid setups where power sources may vary. Proper grounding prevents electrical shocks and helps protect your system from lightning strikes.

- Grounding the Battery Bank: Ensure that your battery bank is properly grounded to prevent dangerous voltage buildups.

- Grounding Electrical Outlets: Each outlet should be grounded to a grounding rod or plate buried outside your home.

Step 4: Installing the Inverter

Most off-grid systems use direct current (DC) from solar panels or wind turbines, but household appliances run on alternating current (AC). An inverter converts DC into AC for household use.

- Location: Install the inverter close to your battery bank to minimize energy loss. Ensure it has proper ventilation to prevent overheating.

- Inverter Size: Choose an inverter that matches the total energy demand of your home, with some room for future expansion.

Voltage Considerations: AC vs. DC Power

One of the unique aspects of off-grid homes is the decision between using alternating current (AC), direct current (DC), or a combination of both. Understanding the difference between AC and DC, and knowing when to use each, can significantly affect the efficiency of your energy distribution system.

1. Understanding AC and DC Power

- Direct Current (DC): DC electricity flows in a constant direction and is typically used in batteries, solar panels, and small appliances. It's more efficient for transmitting power over short distances.

- Alternating Current (AC): AC electricity changes direction periodically and is the standard for household appliances and grid power. It's better for transmitting power over long distances and is more compatible with most modern appliances.

2. Deciding Between AC and DC

Many off-grid systems use DC power from solar panels, wind turbines, or battery banks, but since most household appliances run on AC, an inverter is required to convert DC to AC. However, in some cases, using DC for certain appliances can improve efficiency.

When to Use DC Power

- For Short Distances: If you're running power directly from a battery bank to DC-powered devices (such as LED lighting or low-voltage appliances), using DC can minimize energy loss.

- For Small Loads: DC is ideal for small loads like lighting, fans, or USB-powered devices, especially if you want to avoid the energy loss that occurs during the DC-to-AC conversion process.

When to Use AC Power

- For Household Appliances: Most standard household appliances, such as refrigerators, washing machines, and power tools, require AC power. An inverter is necessary to convert the DC power from your battery bank into usable AC electricity.

- For Long-Distance Power Distribution: If you need to transmit power over long distances (such as from a solar array far from your home), AC power is more efficient and experiences less voltage drop than DC.

3. Hybrid Systems: Combining AC and DC

Many off-grid homes use a hybrid approach, where DC is used for low-voltage devices, while an inverter converts DC to AC for larger household appliances. This system reduces the energy loss that occurs when converting power between AC and DC.

Energy Monitoring and Control: Optimizing Your System

Efficient energy distribution isn't just about wiring and voltage; it also requires ongoing monitoring and control to ensure your system operates smoothly. Energy management systems help you track energy usage, optimize consumption, and make adjustments to prevent outages or overloading.

1. Energy Management Systems: Monitoring Usage

Energy management systems are essential for off-grid living, where energy resources are often limited. These systems allow you to monitor how much energy you're producing, storing, and consuming in real time, helping you make informed decisions about energy usage.

Key Features of an Energy Management System

- Real-Time Monitoring: Energy management systems provide real-time data on your energy production, storage levels, and consumption. You can track how much energy your solar panels, wind turbines, or other sources are generating at any given time.

- Historical Data: Many systems offer historical data tracking, allowing you to see patterns in energy usage over days, weeks, or months. This can help you identify peak consumption periods and adjust your energy production accordingly.

- Alerts and Notifications: Some systems send alerts when your battery levels are low or when energy consumption spikes, helping you avoid overloading your system or running out of stored energy.

Popular Energy Management Tools

- Victron Energy: Offers a range of off-grid monitoring tools that provide detailed insights into your system's performance, including battery levels, solar production, and energy consumption.

- MidNite Solar's Classic: This charge controller comes with an energy monitoring feature that tracks solar panel output and battery storage in real time.

2. Optimizing Energy Usage with Monitoring

Once you have a system in place to monitor your energy usage, you can take steps to optimize how you consume power:

- Shift High-Energy Tasks: Perform energy-intensive tasks, such as laundry or running the dishwasher, during peak energy production hours (such as midday, when solar output is highest).

- Stagger Energy Consumption: Avoid running multiple high-energy appliances at the same time to prevent overloading your system.

- Identify Energy Drains: Use your monitoring system to identify appliances or devices that consume large amounts of energy. Replace inefficient appliances with more energy-efficient models to reduce overall consumption.

Automatic Controls: Enhancing Efficiency and Convenience

Automation can play a significant role in off-grid energy management by allowing you to control when and how energy is used. By using automatic controls, you can reduce waste, improve efficiency, and ensure your system is running optimally without constant manual adjustments.

1. Automating Lights and Appliances

Automation systems can control lights, appliances, and other devices based on energy availability, time of day, or occupancy. These controls reduce unnecessary energy consumption and make your off-grid system more efficient.

Lighting Automation

- Motion Sensors: Installing motion sensors in rooms can ensure that lights turn on only when someone is present, reducing wasted energy.

- Timed Lighting: Use timers to control outdoor or security lighting. Lights can automatically turn on at sunset and off at sunrise, ensuring energy is used only when necessary.

Appliance Automation

- Smart Plugs: Smart plugs allow you to control appliances remotely via a smartphone app or voice command. You can set appliances to turn on or off based on energy availability, preventing them from drawing power when not in use.

- Load Shedding: Some energy management systems include load-shedding features, which automatically turn off non-essential appliances during periods of low energy production. This helps conserve power and prioritize critical devices, like refrigerators or medical equipment.

2. Temperature Controls

Automating your heating and cooling systems ensures that your home stays comfortable without using excess energy. Here's how to manage your home's temperature efficiently:

- Programmable Thermostats: A programmable thermostat adjusts your home's temperature based on your daily routine. You can program the temperature to less when you're away and increase when you return, saving energy during the day.

- Zoned Heating and Cooling: Use automatic controls to manage temperature zones in your home. For example, you can heat only the rooms you're using, rather than the entire house, reducing energy consumption.

3. Backup Generator Automation

Many off-grid homes have backup generators for emergency power. Automating your generator ensures that it only runs when necessary, preserving fuel and reducing wear and tear.

- Automatic Start/Stop: Use an automatic start/stop system to turn the generator on when battery levels drop below a certain threshold or when energy production is insufficient.

- Fuel Monitoring: Install a fuel monitoring system to keep track of your generator's fuel levels and receive alerts when it's time to refuel.

Building a Resilient Off-Grid Energy System

By combining efficient wiring, voltage considerations, energy monitoring, and automatic controls, you can create an off-grid energy distribution system that is both reliable and sustainable. Proper planning and integration of these systems will ensure that your home stays powered, comfortable, and efficient, even during periods of low energy production. Whether you're wiring your home, choosing between AC and DC power, or automating key systems, the right setup will help you make the most of your renewable energy sources and achieve true energy independence.

12. Energy Efficiency, Reducing Consumption and Maximizing Comfort in Off-Grid Living

Living off the grid requires not only generating enough energy but also using that energy efficiently. Maximizing energy efficiency can drastically reduce your overall power consumption, making your renewable energy system more sustainable and cost-effective. In this section, we will explore how choosing low-energy appliances and incorporating insulation and heat retention strategies can help you minimize energy waste while maintaining comfort in your off-grid home.

Low-Energy Appliances: Cutting Power Consumption at the Source

One of the most impactful methods to enhance energy efficiency is by selecting low-energy appliances. These appliances are specifically designed to consume less power without sacrificing performance. Whether it's for cooking, refrigeration, lighting, or heating, opting for energy-efficient models reduces the strain on your energy system and helps you achieve greater independence from traditional power sources.

1. The Importance of Choosing Energy-Efficient Appliances

Off-grid homes often operate on limited energy supplies from solar panels, wind turbines, or other renewable sources. Therefore, every kilowatt-hour saved by using efficient appliances contributes to your overall energy sustainability. Here are some key reasons to prioritize energy-efficient appliances:

- Reduced Energy Demand: Low-energy appliances require less power, allowing you to store more energy for other uses, such as lighting, heating, or charging batteries.

- Smaller Energy Systems: With lower energy consumption, you can size your renewable energy system smaller, potentially reducing the number of solar panels, batteries, or wind turbines needed.

- Lower Costs: Though energy-efficient appliances may have a higher upfront cost, they save money in the long run by reducing your energy bills (or prolonging the life of your battery bank in an off-grid setting).

2. Key Features of Low-Energy Appliances

When choosing energy-efficient appliances for your off-grid home, look for specific features that minimize power consumption without sacrificing performance:

- Energy Star Certification: Appliances with an Energy Star label meet government-backed standards for energy efficiency, ensuring they consume less electricity than non-certified models.

- Inverter Technology: Inverter appliances, particularly in refrigerators and air conditioners, adjust their speed based on demand, consuming only the necessary power. This reduces the frequent on-and-off cycling that causes energy spikes.

- Low Standby Power: Many appliances continue to draw power even when turned off. Look for models with low standby power consumption (less than 1 watt), or use smart plugs to completely shut off power when appliances are not in use.

- Programmable and Smart Features: Appliances with programmable or smart features can be scheduled to run during peak energy production hours (e.g., midday for solar power), reducing demand on your battery bank at night.

3. Choosing Energy-Efficient Appliances for Your Off-Grid Home

Let's explore some common household appliances and how to select the most energy-efficient models for off-grid living.

Energy-Efficient Refrigerators

Refrigerators are one of the largest energy consumers in any household. Choosing a low-energy refrigerator is critical for maintaining food storage without overloading your energy system.

- Look for Top-Freezer Models: Top-freezer refrigerators are generally more energy-efficient than side-by-side or bottom-freezer models because the freezer compartment benefits from the cold air naturally settling in the lower part of the fridge.

- Consider a DC-Powered Refrigerator: If your off-grid system uses direct current (DC) power from solar panels or wind turbines, opt for a DC-powered refrigerator to avoid energy loss during the AC-to-DC conversion process. DC fridges are designed for off-grid use and tend to be more efficient.

- Insulation: Choose a refrigerator with thick insulation, as this reduces the amount of time the compressor needs to run to maintain cold temperatures. The thicker the insulation, the more energy-efficient the unit.

Practical Tip: If you have an older refrigerator that isn't energy-efficient, consider adding additional insulation or installing a temperature-controlled timer to reduce its energy use during the night.

Energy-Efficient Lighting

Lighting is another key area where energy savings can be made. Modern lighting solutions are far more efficient than older incandescent bulbs.

- LED Bulbs: Light-emitting diode (LED) bulbs are the most energy-efficient lighting option available. They use up to 90% less energy than traditional incandescent bulbs and last significantly longer.

- Solar-Powered Lighting: For outdoor lighting, consider using solar-powered lights. These lights store energy during the day and automatically turn on at night, eliminating the need for wired electricity.

- Task Lighting: Instead of lighting an entire room, use task lighting (e.g., desk lamps or under-cabinet lighting) for focused activities. This reduces overall energy consumption and ensures that light is only used where it's needed.

Energy-Efficient Cooking Appliances

Cooking can require a significant amount of energy, especially for off-grid homes that rely on electric stoves or ovens. Choosing the right energy-efficient cooking appliances can save both electricity and time.

- Induction Cooktops: Induction cooktops are highly energy-efficient because they use electromagnetic fields to directly heat the cookware, rather than the cooktop itself. This results in faster cooking times and less energy waste.

- Solar Ovens: In sunny climates, solar ovens are a fantastic energy-saving alternative. These ovens use the sun's energy to bake, roast, or steam food without using any electricity.

- Pressure Cookers: Pressure cookers cook food faster than traditional stovetop methods by trapping steam and increasing pressure inside the pot. This reduces cooking time and energy consumption.

Practical Tip: When using an electric stove, match the size of the cookware to the burner to minimize heat loss and energy waste.

Energy-Efficient Heating and Cooling

Heating and cooling can be some of the largest energy drains in any home. For off-grid living, efficient systems are essential to reduce energy consumption and maintain comfort.

- Heat Pumps: A heat pump is one of the most energy-efficient ways to heat and cool a home. It transfers heat from the outside air (even in cold weather) into your home, using significantly less energy than conventional heating systems.

- Solar Water Heaters: Solar water heaters use energy from the sun to heat water, which can be used for showers, cleaning, or radiant floor heating. These systems can drastically reduce the need for electric or gas-powered water heaters.

- Ceiling Fans: Ceiling fans use very little electricity and can help circulate warm or cool air more efficiently, reducing the need for air conditioning or heating systems.

Insulation and Heat Retention: Keeping Energy Where It Belongs

No matter how energy-efficient your appliances are, you'll waste energy if your home isn't properly insulated. Insulation and heat retention strategies are vital to reducing energy loss, keeping your home warm in the winter and cool in the summer without relying heavily on heating or cooling systems.

1. The Role of Insulation in Off-Grid Homes

Insulation works by slowing the transfer of heat between the inside and outside of your home. The better insulated your home is, the less energy you'll need to maintain a comfortable indoor temperature.

- Reducing Heating and Cooling Costs: Proper insulation keeps warm air inside during the winter and prevents cool air from escaping in the summer. This reduces the need for heating and air conditioning, saving energy.

- Comfort and Indoor Air Quality: Insulation not only improves energy efficiency but also enhances comfort by maintaining consistent indoor temperatures. It can also help reduce noise pollution and prevent drafts.

2. Types of Insulation Materials

There are various types of insulation materials to choose from, each with different properties suited to specific areas of your home. Here's a look at the most common types of insulation:

Fiberglass Insulation

Fiberglass is one of the most widely used insulation materials. It's made from fine strands of glass and is available in batt or loose-fill form.

- Best For: Walls, floors, and attics.

- Advantages: Fiberglass is affordable, non-combustible, and resistant to moisture damage.

- Limitations: It requires protective gear for installation, as the tiny glass fibers can irritate the skin and lungs.

Spray Foam Insulation

Spray foam is a high-performance insulation material that expands to fill gaps and cracks, providing an airtight seal. It's available in two forms: open-cell and closed-cell.

- Best For: Walls, floors, attics, and areas with difficult-to-reach gaps.

- Advantages: Spray foam offers excellent thermal resistance and air-sealing properties. Closed-cell foam also adds structural strength.

- Limitations: It's more expensive than other types of insulation and must be installed by professionals.

Cellulose Insulation

Cellulose insulation is made from recycled paper products and is treated with fire-retardant chemicals to improve safety.

- Best For: Walls and attics, particularly for retrofitting older homes.

- Advantages: Cellulose is eco-friendly, affordable, and provides good thermal performance.

- Limitations: It can settle over time, reducing its effectiveness, and may absorb moisture if not installed properly.

Reflective Insulation

Reflective insulation (also known as radiant barriers) is designed to reflect heat rather than absorb it. It's typically made from aluminum foil and is installed in attics or walls.

- Best For: Hot climates where cooling costs are a concern.

- Advantages: Reflective insulation reduces heat gain during the summer by reflecting radiant heat away from your home.

- Limitations: It's less effective in cold climates, as it doesn't provide significant insulation against heat loss.

Insulation Techniques for Maximum Heat Retention

Installing insulation is just the first step. To maximize heat retention, you need to implement smart design strategies and construction techniques that prevent energy loss.

1. Insulating Walls and Ceilings

Properly insulating your home's walls and ceilings is critical to maintaining comfortable indoor temperatures without excessive energy use.

- Double-Stud Walls: Double-stud walls are a common design for energy-efficient homes. By creating a thicker wall cavity, you can add more insulation, increasing the home's thermal resistance.

- Blown-In Insulation: Blown-in insulation is ideal for filling gaps in walls or ceilings, particularly in older homes where traditional batts may not fit properly. It's blown into the wall cavities, filling every nook and cranny to create a continuous thermal barrier.

Practical Tip: Insulate interior walls between heated and unheated spaces (such as garages or storage rooms) to prevent heat loss into those areas.

2. Sealing Air Leaks

Air leaks are one of the biggest contributors to energy loss in homes. Even with good insulation, cracks, gaps, and leaks around windows, doors, and pipes can allow warm or cool air to escape, forcing your heating or cooling system to work harder.

- Caulking and Weatherstripping: Use caulk to seal small gaps around windows and doors, and install weatherstripping to create a tight seal when doors and windows are closed.

- Expanding Foam: For larger gaps around pipes, vents, or in the attic, expanding spray foam is ideal for filling irregular spaces and preventing air leaks.

3. Insulating the Roof and Attic

A significant amount of heat is lost through the roof and attic. Ensuring that these areas are properly insulated will help keep your home warmer in the winter and cooler in the summer.

- Roof Insulation: Insulate the underside of the roof to reduce heat transfer. In colder climates, choose insulation with a high R-value (a measure of thermal resistance) to retain heat.

- Attic Ventilation: While insulation is important, ventilation is also crucial for preventing moisture buildup in the attic. Install soffit vents and ridge vents to maintain airflow and reduce the risk of condensation, which can damage insulation.

4. Using Energy-Efficient Windows

Windows are another major source of heat loss. Energy-efficient windows, combined with proper sealing techniques, can significantly reduce this loss.

- Double-Glazed Windows: Double-glazed windows feature two panes of glass with an insulating layer of gas (usually argon) in between. This aids in minimizing heat transfer and prevent drafts.

- Low-E Coatings: Windows with low-emissivity (Low-E) coatings reflect infrared heat back into your home, keeping it warmer in the winter and cooler in the summer.

Passive Solar Design: Harnessing Natural Energy for Heating and Cooling

Passive solar design involves using the sun's energy to heat or cool your home naturally, reducing the need for artificial heating and cooling systems. By incorporating smart design principles, you can maximize solar gain during the winter and minimize it during the summer.

1. Orienting Your Home for Maximum Solar Gain

The orientation of your home plays a critical role in how much sunlight it receives, especially in the winter when solar gain is most important.

- South-Facing Windows: In the Northern Hemisphere, orient your home so that the largest windows face south. This allows you to capture the maximum amount of sunlight during the day, reducing the need for heating.

- Roof Overhangs: Install roof overhangs to block the sun during the summer when it's higher in the sky, while still allowing sunlight to enter during the winter months.

2. Thermal Mass for Heat Storage

Thermal mass refers to materials that absorb and store heat during the day and release it slowly at night. Incorporating thermal mass into your home's design can help regulate indoor temperatures and reduce energy consumption.

- Concrete Floors: Concrete is an excellent thermal mass material. When sunlight hits the floor, it absorbs heat and radiates it back into the room after the sun goes down.

- Masonry Walls: Thick masonry walls (made from materials like brick or stone) also store heat during the day and release it slowly at night, helping to maintain a stable indoor temperature.

3. Natural Ventilation for Cooling

In addition to heating, passive solar design can also be used for cooling by incorporating natural ventilation strategies.

- Cross Ventilation: Cross ventilation allows cool air to flow through your home, pushing hot air out. By placing windows on opposite sides of the house, you can create a natural airflow that cools your home without the need for air conditioning.

- Stack Ventilation: In homes with high ceilings, warm air naturally rises to the top of the room. Stack ventilation uses this principle by installing vents or windows at the top of the house to allow hot air to escape, keeping the lower levels cooler.

Conclusion: Achieving Maximum Energy Efficiency in Your Off-Grid Home

Achieving energy efficiency in an off-grid home is a multi-faceted process that requires careful consideration of both appliance selection and building design. By choosing low-energy appliances, properly insulating your home, and incorporating passive solar design strategies, you can minimize energy loss and reduce overall power consumption. These steps not only make your home more comfortable and sustainable but also ensure that your renewable energy system is able to meet your needs year-round.

Module D | Off-Grid Waste Management

13. Composting Toilets, Sustainable Waste Management for Off-Grid Living

When living off the grid, managing waste effectively is critical not only for hygiene but also for sustainability. Composting toilets offer a solution that eliminates the need for water-based sewage systems while turning human waste into usable compost for gardens. In this section, we'll explore the different types of composting toilets, how to maintain them, and best practices for safe composting.

Understanding Composting Toilets: A Practical Overview

A composting toilet is designed to break down human waste into compost through natural decomposition, reducing waste and eliminating the need for septic systems or sewage connections. These systems use aerobic bacteria, oxygen, and time to break down waste into a soil-like substance. Here's why composting toilets are perfect for off-grid living:

- Water Conservation: Unlike traditional toilets, composting toilets don't use water, which is crucial for off-grid homes where water resources may be limited.

- Simplicity: Composting toilets can be easy to set up and maintain, requiring no plumbing or complicated infrastructure.

- Environmental Impact: By turning waste into compost, you're not only reducing pollution but also contributing to soil fertility if the compost is used safely.

1. Types of Composting Toilets

Composting toilets come in several designs, each suited for different needs and levels of use. The key is to find the right system that aligns with your off-grid setup, budget, and lifestyle.

A. Self-Contained Composting Toilets

Self-contained composting toilets are compact, all-in-one units that are easy to install and maintain. They are perfect for small off-grid homes, cabins, or tiny houses with limited space. These units handle both waste processing and composting in the same unit.

- How They Work: Waste is deposited directly into a compartment within the toilet, where it is mixed with a bulking agent (like sawdust or peat moss) to aid decomposition and control odor. The composting chamber is vented to allow for airflow, and some models include fans or heating elements to accelerate the composting process.

- Maintenance: Self-contained units need regular emptying—usually every few months, depending on usage. The compost must be carefully handled and allowed to further decompose in a designated composting area before being used as fertilizer.

- Best For: Tiny homes, small cabins, or locations with minimal space.

Example: The Nature's Head composting toilet is a popular self-contained option, known for its ease of installation and minimal odor control needs.

B. Central Composting Systems

Central composting toilets involve a separate composting unit, usually located in a basement or outdoors. The toilet itself may look like a conventional toilet, but instead of sending waste to a septic tank, it directs it to the central composting chamber via a pipe.

- How They Work: Waste is flushed (either with a small amount of water or using gravity) to a central composting bin, where it decomposes over time. These systems often have larger composting chambers, allowing them to handle higher volumes of waste.

- Maintenance: Because of the larger composting capacity, central systems can go longer between maintenance cycles. However, you'll need to monitor moisture levels and occasionally mix the compost for even decomposition.

- Best For: Off-grid homes with more space or families with multiple users.

Example: The Sun-Mar Centrex system offers a central composting design with various models that can handle different user capacities and climates.

C. Urine-Diverting Composting Toilets

Urine-diverting composting toilets separate liquid waste (urine) from solid waste. Since urine contains the majority of nitrogen, separating it from solid waste helps reduce odors and makes the composting process faster and more efficient.

- How They Work: These toilets have two chambers: one for liquid waste and another for solids. The urine is either stored separately for later use as a fertilizer (after dilution) or disposed of, while solid waste is treated like traditional compost.

- Maintenance: You'll need to empty the liquid waste container regularly and manage the solid waste chamber similarly to self-contained composting toilets.

- Best For: Homes in hot climates where odor control is more challenging, or for users looking for quicker composting cycles.

Example: The Separett Villa 9215 is a popular urine-diverting toilet that's easy to maintain and offers a higher level of odor control.

2. Key Factors for Choosing a Composting Toilet

Choosing the right composting toilet depends on several factors, including the number of users, climate, and available space. Here's what you need to consider before making your choice:

A. User Capacity

Different composting toilets are designed for different user capacities. Some units are built for one or two people, while larger models can accommodate families or even small communities. Be sure to choose a toilet that matches your household size to avoid excessive maintenance or overloading the composting chamber.

- For Single Users or Couples: A self-contained unit will likely meet your needs.

- For Families: Opt for a central composting system or a larger-capacity self-contained toilet to reduce the frequency of emptying.

B. Climate Considerations

The climate of your off-grid location can affect how well your composting toilet functions. Warmer climates accelerate the composting process, while colder regions may slow it down, requiring additional steps like insulation or heating.

- Warm Climates: In warmer areas, composting happens faster, but you may need to take extra precautions to manage odors and moisture levels.

- Cold Climates: In colder climates, a composting toilet may need added insulation or a heating element to maintain the right temperature for decomposition.

C. Space and Location

Think about where the toilet will be installed and how much space you have available. For tiny homes or cabins, a self-contained unit may be the easiest option. If you have more room, a central composting system or an outdoor composting chamber could be more practical.

3. Maintaining a Composting Toilet

Once you've chosen the right composting toilet for your needs, it's essential to maintain it properly to ensure long-term functionality and safety. Here are some key maintenance tasks for composting toilets:

A. Adding Bulking Material

To aid in the decomposition process and control odors, you'll need to add a bulking material (such as sawdust, coconut coir, or peat moss) after each use. This material absorbs moisture, adds carbon, and helps keep the compost aerated.

- How Much to Add: Typically, you'll add a small scoop of bulking material after each use. This helps cover the waste, reducing odors and promoting better breakdown.

- Choosing the Right Material: Make sure the bulking material you use is dry and carbon-rich, which will help balance the nitrogen from the waste.

B. Monitoring Moisture Levels

Composting toilets need the right balance of moisture to function properly. Too much moisture can lead to anaerobic conditions (which slow decomposition and create foul odors), while too little moisture can halt the composting process altogether.

- Ideal Moisture Level: The compost should feel like a wrung-out sponge—damp but not wet. If the compost is too wet, add more bulking material or increase ventilation. If it's too dry, you can add a small amount of water or adjust the humidity in the composting chamber.

C. Managing Odors

While a well-maintained composting toilet should not produce significant odors, there are steps you can take to ensure optimal odor control:

- Ventilation: Many composting toilets come with a venting system to expel odors. Ensure the vent is properly installed and clear of blockages.

- Add Carbon Material: Adding enough bulking material (carbon) after each use helps reduce odors by balancing out the nitrogen from waste.

- Regular Stirring: For some units, regularly stirring the compost helps aerate it, promoting faster breakdown and reducing smells.

D. Emptying the Compost

Depending on the size and type of your composting toilet, you'll need to empty the compost periodically. For self-contained units, this may be every few months, while central systems can go longer between emptying.

- How to Empty: When the composting chamber is full, you'll need to transfer the partially composted material to an outdoor compost bin for further decomposition. Be sure to wear gloves and use proper sanitation practices when handling human waste.

- Post-Emptying Decomposition: It's important to allow the compost to cure for at least six months to a year before using it on non-edible plants. This ensures that any pathogens are fully broken down.

4. Safe Composting Practices

While composting toilets are an eco-friendly way to handle waste, it's essential to follow safety guidelines to prevent contamination and ensure the compost is safe to use.

A. Handling Human Waste Safely

Human waste can contain harmful pathogens, so it's important to handle it with care, especially during the composting process. Always wear gloves when handling composted material and wash your hands thoroughly afterward.

- Secondary Composting: After removing compost from the toilet, it's recommended to continue the composting process in a separate bin for at least 6-12 months to ensure complete pathogen breakdown. During this time, keep the compost covered and away from direct contact with food plants.

B. Using Compost in the Garden

The compost generated from human waste can be used to enrich soil, but it's generally recommended to avoid using it on food crops. Instead, use it for trees, shrubs, or other ornamental plants.

- Non-Edible Use: Apply the compost to non-edible plants such as fruit trees or flower gardens. This adds nutrients to the soil while ensuring food safety.

- Follow Local Regulations: Be sure to check local regulations regarding the use of composted human waste, as some areas have strict guidelines on how and where it can be used.

5. Troubleshooting Common Issues

Even with proper maintenance, you may encounter some issues with your composting toilet. Here are a few common problems and how to address them:

A. Strong Odors

If you notice strong odors coming from your composting toilet, it could be a sign of insufficient ventilation, too much moisture, or an imbalance in the composting process.

- Solutions: Check the vent for blockages, add more carbon material, and ensure the compost isn't too wet. Stirring the compost to aerate it can also help reduce odors.

B. Slow Composting

If your compost isn't breaking down as quickly as expected, it could be due to a lack of heat, oxygen, or moisture.

- Solutions: Ensure the composting chamber has proper airflow and maintain the ideal moisture level. In colder climates, you may need to add a heating element to speed up the decomposition process.

C. Overflow or Excess Moisture

If your composting toilet becomes too full or the compost is too wet, it can lead to overflow issues or improper composting.

- Solutions: Empty the composting chamber more frequently, or adjust the amount of bulking material you're adding. For central systems, check that the drainage system is working properly.

14. Greywater Recycling, Safe and Sustainable Water Reuse for Off-Grid Living

Water is one of the most precious resources in an off-grid setting, and maximizing its use through recycling can make a significant difference in your overall water sustainability. Greywater recycling is an efficient and environmentally friendly way to reuse water from sinks, showers, and washing machines for non-potable purposes like irrigation. This not only conserves fresh water but also reduces the burden on your septic or waste management system. In this section, we'll cover various methods for safely recycling greywater, ensuring it's clean enough for plant irrigation without harming the environment.

What Is Greywater?

Greywater is the relatively clean wastewater generated from household activities like washing dishes, taking showers, and doing laundry. Unlike blackwater, which comes from toilets and contains harmful pathogens, greywater can be safely reused for irrigation or landscaping with minimal treatment

Sources of Greywater:

- Bathroom sinks and showers
- Washing machines
- Kitchen sinks (if food and grease content are minimized)

Greywater typically contains soap, dirt, and organic matter, but when properly managed, it can be a valuable resource for off-grid homes.

1. Benefits of Greywater Recycling

Recycling greywater offers several advantages, especially for off-grid homes where water conservation is crucial:

- Water Conservation: By reusing water that would otherwise go to waste, you reduce the demand on your primary water source, such as rainwater or well water.

- Reduced Strain on Septic Systems: Greywater recycling reduces the amount of water flowing into your septic tank, prolonging its lifespan and lowering maintenance costs.

- Lower Water Bills: For those with access to a water utility, recycling greywater can significantly reduce water bills by minimizing freshwater use for irrigation.

- Environmentally Friendly: Greywater recycling minimizes wastewater output and helps reduce the overall environmental impact of your household.

2. Key Considerations for Safe Greywater Recycling

While greywater recycling is a sustainable practice, it's important to follow safety guidelines to prevent contamination and ensure the water is safe for your plants and the environment.

A. Understanding Water Quality

Greywater can vary in quality depending on its source. For example, water from a bathroom sink or shower may contain soaps and shampoos, while water from a washing machine may have detergent residue. You'll need to consider the quality of the greywater you're recycling and treat it appropriately to avoid harming your plants.

- Avoid Toxic Substances: Be mindful of the products you use in your home. Choose biodegradable soaps, detergents, and cleaning products to ensure your greywater is safe for reuse in the garden.

- Minimize Solids and Grease: If you plan to use greywater from kitchen sinks, avoid letting food particles, grease, or oils enter the system, as these can clog filters and harm plants.

B. Local Regulations

Before installing a greywater system, it's important to check local regulations, as some areas have specific guidelines on how greywater can be reused. Ensure that your system complies with local health and safety standards to avoid any legal or environmental issues.

3. Greywater Recycling Systems

There are several ways to collect and recycle greywater, ranging from simple DIY setups to more complex systems that filter and distribute water automatically. Let's explore some common greywater recycling methods suitable for off-grid living.

A. Gravity-Fed Systems: Simple and Effective

Gravity-fed greywater systems are the simplest and most cost-effective method of recycling water for irrigation. These systems rely on the natural force of gravity to move water from sinks or showers to a designated irrigation area in your garden.

- How It Works: Greywater flows directly from the source (e.g., bathroom sink or shower) into a collection pipe that leads to a garden or landscaping area. The system uses gravity to transport the water, eliminating the need for pumps or electricity.

- Best For: Small homes or gardens with a slight slope that allows water to flow naturally downhill.

Installation Steps:

1. Collect the Greywater: Install a diverter valve under the sink or shower drain to collect greywater. This allows you to switch between sending water to the septic system or to your garden.

2. Install Piping: Use flexible hoses or PVC pipes to direct the greywater from the collection point to your garden. Ensure the pipes are sloped downward to allow gravity to do the work.

3. Create an Irrigation Zone: Designate an area in your garden where greywater will be dispersed. This can be a mulch basin or an area with water-tolerant plants that thrive on greywater.

Practical Tip: Use coarse gravel or mulch in the irrigation area to filter the water further and prevent pooling, which could lead to odors or attract pests.

B. Branched Drain Systems: Efficient Distribution for Larger Gardens

For off-grid homes with larger gardens or landscapes, a branched drain system offers a more advanced greywater recycling solution. These systems use a series of pipes and valves to direct greywater to multiple irrigation zones, ensuring that water is evenly distributed across your garden.

- How It Works: Greywater is collected from one or more sources and directed through a series of branched pipes that lead to different areas of your garden. Each branch can be individually controlled, allowing you to direct water where it's needed most.

- Best For: Larger properties with multiple irrigation zones or gardens that need more even water distribution.

Installation Steps:

1. Collect Greywater: As with gravity-fed systems, install a diverter valve to collect greywater from sinks, showers, or washing machines.

2. Set Up Branch Pipes: Use PVC pipes to create a network of branches that lead to different parts of your garden. Each branch should be angled to allow water to flow evenly.

3. Install Valves: Install manual or automatic valves on each branch to control the flow of water to specific irrigation zones. This allows you to prioritize certain plants based on their water needs.

Practical Tip: Install a sediment trap at the beginning of the system to prevent solids from clogging the pipes and valves. Clean the trap regularly to ensure smooth operation.

C. Greywater Filters and Treatment Systems

In some cases, particularly when irrigating food crops, it's necessary to treat greywater to remove contaminants and ensure it's safe for use in your garden. Greywater filters and treatment systems provide additional filtration, allowing you to use greywater for more sensitive plants or even in limited household applications.

- How It Works: These systems use a series of filters to remove solids, soap residues, and other contaminants from greywater before it's reused. Filters can range from simple mesh screens to more advanced biological filters that use plants and microorganisms to clean the water.

- Best For: Homes where greywater will be used for more delicate plants or where regulations require additional filtration.

Types of Greywater Filters:

1. Mesh Filters: Basic mesh filters are installed at the collection point to trap larger solids, such as hair or food particles. These filters require regular cleaning but are simple and inexpensive.

2. Sand Filters: Sand filters use layers of fine sand and gravel to filter greywater, removing smaller particles and improving water clarity. These are ideal for greywater systems used in larger gardens or orchards.

3. Constructed Wetlands: Constructed wetlands are a natural form of greywater treatment, using plants and microorganisms to break down contaminants. Greywater flows through a series of gravel-filled beds planted with wetland vegetation, which filters and cleans the water.

Practical Tip: Use biodegradable soaps and detergents to reduce the number of contaminants that need to be filtered from your greywater, extending the lifespan of your filtration system.

4. Using Greywater Safely in the Garden

While greywater is generally safe for irrigation, there are some important considerations to ensure that it doesn't harm your plants or soil. Properly managing how and where greywater is applied can prevent issues like nutrient overload, root damage, or contamination.

A. Selecting Plants for Greywater Irrigation

Not all plants are well-suited for greywater irrigation, especially those that are sensitive to soap or detergent residues. Here are some plant types that thrive on greywater:

- Fruit Trees: Trees such as apple, peach, and citrus are excellent candidates for greywater irrigation, as they require more water and are less sensitive to the mild detergents present in greywater.

- Non-Edible Plants: Shrubs, flowers, and ornamental plants generally tolerate greywater well, especially when it's applied through a mulch basin or drip system.

- Drought-Tolerant Plants: Plants adapted to dry conditions, such as lavender, rosemary, or succulents, benefit from greywater irrigation, especially in areas with limited rainfall.

B. Avoiding Contact with Edible Plants

While greywater can be safely used for irrigation, it's recommended to avoid using it directly on edible plants, particularly root vegetables or leafy greens that are eaten raw. If you want to use greywater in your vegetable garden, make sure it doesn't come into direct contact with the plants.

- Irrigate Below the Surface: Use drip irrigation or subsurface methods to deliver greywater directly to the root zone, reducing the risk of contamination.

- Keep Greywater Away from Edible Parts: For crops like tomatoes or peppers, ensure that greywater is applied to the base of the plants and doesn't splash onto leaves or fruit.

C. Preventing Soil and Plant Damage

Over time, greywater can accumulate salts or other compounds from soaps and detergents, which may harm sensitive plants or soil health. To mitigate these risks, consider the following practices:

- Rotate Irrigation Areas: To prevent nutrient buildup or salt accumulation, rotate greywater irrigation areas throughout your garden.

- Flush Soil with Freshwater: Periodically irrigate greywater areas with freshwater to dilute any salt or soap residues and prevent soil degradation.

5. Maintaining a Greywater System

To keep your greywater system functioning effectively, regular maintenance is essential. Here's what you need to do to ensure your system stays clean and efficient:

A. Cleaning Filters and Traps

If your system uses filters or sediment traps, clean them regularly to prevent blockages and ensure smooth water flow.

- How Often: Depending on usage, clean filters every month to prevent clogs. For mesh or sand filters, inspect and clean them more frequently if you notice a decrease in water flow.

- Replacement: Replace filters as needed, especially if they show signs of wear or reduced effectiveness.

B. Inspecting Piping and Valves

Check your piping and valves periodically to ensure there are no leaks or blockages in the system.

- Look for Signs of Blockages: If water flow slows down, inspect the pipes for buildup or clogs and clean them out as needed.

- Check for Leaks: Ensure that all connections and joints are secure and that there are no leaks that could waste water or damage plants.

6. Troubleshooting Common Greywater Issues

Like any system, greywater recycling may present challenges from time to time. Here are some common issues and how to resolve them:

A. Clogged Pipes or Filters

If greywater is not flowing smoothly, the most likely cause is a clogged pipe or filter.

- Solution: Clean the filters and check for blockages in the pipes. If the clog persists, use a flexible cleaning rod to clear out any debris.

B. Foul Odors

While properly managed greywater systems shouldn't produce odors, improper drainage or standing water can cause foul smells.

- Solution: Ensure that your irrigation zones are draining properly and that greywater is being dispersed evenly across your garden. If odors persist, check the filtration system for blockages or excess moisture.

C. Poor Plant Growth

If your plants seem to be struggling despite regular irrigation with greywater, the cause may be related to soap or nutrient buildup.

- Solution: Flush the soil with freshwater to remove any buildup of salts or soap residues. Consider rotating irrigation areas or using a more advanced filtration system if the problem continues.

15. Humanure Systems, Safely Composting Human Waste for Garden Use

Humanure systems are a sustainable way to manage human waste off the grid by turning it into valuable compost that can be safely used in gardens. While the concept of composting human waste may seem unconventional, it's an eco-friendly approach that reduces waste, conserves water, and enriches the soil. However, to ensure the process is safe and sanitary, specific methods and precautions must be followed. In this section, we will explore how humanure systems work, the steps for proper composting, and best practices to ensure safe handling and use in your garden.

What is a Humanure System?

A humanure system is a composting process designed specifically for human waste. It uses natural decomposition, aided by microorganisms, to break down waste into nutrient-rich compost. This system differs from traditional composting toilets because it focuses on the complete composting process—from collection to final use in the garden.

- Sustainable Waste Management: Humanure systems allow you to manage waste without the need for water-based plumbing or septic systems, making them ideal for off-grid homes.

- Environmental Benefits: Composting human waste reduces the need for chemical fertilizers and prevents harmful pathogens from contaminating water sources through improper waste disposal.

1. How Humanure Systems Work

Humanure composting involves the collection of human waste in a designated container, followed by its decomposition through aerobic composting. When properly managed, the process safely eliminates harmful pathogens and results in compost that can be used to enrich soil.

The system relies on a few key elements to ensure that the composting process is safe and efficient:

- Collection Bin: A large, sturdy bin or composting toilet serves as the initial collection point for human waste. This bin should have proper ventilation to allow airflow and promote aerobic decomposition.

- Cover Material: To aid decomposition and control odors, each addition of waste is covered with a carbon-rich material, such as sawdust, straw, or leaves. This balances the nitrogen from the waste and keeps the system aerobic.

- Composting Pile or Bin: Once the collection bin is full, the waste is transferred to a larger compost pile or bin, where it undergoes further decomposition. This bin is usually insulated to retain heat, which is essential for killing pathogens.

- Thermal Decomposition: Heat generated during the composting process is a key factor in breaking down harmful bacteria and pathogens, rendering the compost safe for use in gardens.

2. Setting Up a Humanure System

Creating a safe and efficient humanure system for your off-grid home requires careful planning and setup. Here's how to set up your system from start to finish.

A. Building or Choosing a Humanure Toilet

The first step in setting up a humanure system is choosing or building a composting toilet that will serve as your collection point.

- DIY Humanure Toilet: A simple DIY humanure toilet can be built using a 5-gallon bucket and a wooden or plastic toilet seat. The bucket serves as the collection bin, and each use is covered with sawdust or another carbon-rich material to aid decomposition.

- Prefabricated Composting Toilets: Alternatively, you can purchase a prefabricated composting toilet designed for humanure systems. These models often include ventilation systems and are designed to handle waste more efficiently.

Practical Tip: Place the toilet in a well-ventilated area, either inside your home or in an outdoor composting shelter, to control odors and ensure proper airflow.

B. Creating a Composting Area

Once the humanure toilet's collection bin is full, the waste must be transferred to a composting pile or bin. This area should be designed for safe composting, with sufficient space and proper containment.

- Composting Bin: Build or purchase a sturdy composting bin with adequate insulation. The bin should be large enough to allow the compost to reach temperatures of at least 140°F (60°C) to kill pathogens.

- Insulation: Insulate the composting pile with straw, hay, or wood chips to retain heat. The heat generated during decomposition is essential for breaking down harmful organisms.

- Drainage: Ensure that the composting area has good drainage to prevent excess moisture buildup, which could lead to anaerobic (oxygen-free) conditions. This slows decomposition and can cause odors.

3. Safe Composting Practices

Proper handling and composting are crucial to ensuring that humanure systems remain safe and effective. The following best practices help maintain the health of the compost pile and ensure the final product is safe for use.

A. Cover Each Addition with Carbon Material

After each use, cover the waste in the collection bin with a layer of carbon-rich material, such as sawdust, straw, or leaves. This helps balance the carbon-to-nitrogen ratio and controls moisture and odors.

- How Much to Use: A handful of cover material after each use is usually sufficient to maintain the balance and prevent odors.

- Best Materials: Dry, carbon-rich materials like sawdust, shredded newspaper, straw, or dried leaves are ideal for covering waste and promoting aerobic conditions.

B. Monitor Temperature and Moisture Levels

The compost pile needs to reach a high enough temperature to kill pathogens, and it must maintain the right moisture level for optimal decomposition.

- Thermometer: Use a compost thermometer to regularly check the temperature of the pile. For safe composting, the pile should reach and maintain temperatures of 140°F (60°C) or higher for several days.

- Moisture: The compost should feel like a wrung-out sponge—not too dry and not too wet. If it's too dry, add more water or moist materials like grass clippings. If it's too wet, add dry cover material like straw or sawdust to absorb excess moisture.

C. Turn the Pile Regularly

Turning the compost pile aerates it, allowing oxygen to penetrate the pile and promote faster decomposition. Regular turning also helps distribute heat evenly, ensuring that all parts of the pile reach pathogen-killing temperatures.

- When to Turn: Turn the pile every few weeks to ensure proper airflow and to mix the materials evenly. Be sure to wear gloves and follow proper hygiene practices when handling the compost.

4. Ensuring Pathogen Destruction

The primary concern when composting human waste is ensuring that all harmful pathogens are destroyed. Pathogen destruction is achieved through high temperatures and proper composting practices.

A. The Role of Heat in Pathogen Destruction

For humanure compost to be safe, it needs to maintain high temperatures long enough to destroy harmful bacteria, viruses, and parasites.

- Target Temperature: A temperature of at least 140°F (60°C) is required to effectively kill most pathogens. This temperature should be maintained for several days, with the interior of the pile reaching similar temperatures.

B. The Importance of Secondary Composting

Even after the initial composting process is complete, it's recommended to let the compost cure for an additional 6 to 12 months before using it in your garden. This secondary composting phase ensures that any remaining pathogens are destroyed, and the compost fully stabilizes.

- Curing Time: After transferring the compost to the secondary composting bin, allow it to cure undisturbed for at least six months to a year. This ensures the compost is fully matured and safe to use.

5. Using Humanure Compost Safely in the Garden

Once your humanure compost has completed both the primary and secondary composting stages, it's safe to use in your garden. However, some precautions should be taken when applying it to plants, especially edible crops.

A. Use on Non-Edible Plants

The safest and most common use of humanure compost is for non-edible plants such as fruit trees, ornamental shrubs, or flower beds. These plants benefit from the rich nutrients in the compost, while the risk of contamination is minimized since the compost doesn't come into direct contact with food.

- Best Use: Apply humanure compost to trees, shrubs, and other plants where the compost remains on the soil surface or in the root zone.

B. Precautions for Edible Plants

If you decide to use humanure compost on edible plants, especially those that grow in the soil (like root vegetables), take extra precautions to ensure the compost is fully decomposed and pathogen-free.

- Subsurface Application: Apply the compost below the surface, avoiding direct contact with edible parts of the plants. This reduces the risk of contamination.

- Timing: Apply the compost well in advance of planting, ideally at the start of the growing season, to allow further decomposition and reduce any potential pathogen risks.

Practical Tip: Even when fully composted, it's a good idea to avoid using humanure compost on leafy greens or other crops that are eaten raw.

6. Troubleshooting Common Issues in Humanure Systems

While humanure systems are relatively simple to maintain, occasional issues may arise. Here are some common problems and how to address them.

A. Strong Odors

Odors are often a sign that the compost pile is too wet or isn't receiving enough oxygen.

- Solution: Add more carbon-rich cover material, such as sawdust or straw, and ensure the pile is well-ventilated. Turning the pile more frequently can also help reduce odors.

B. Slow Decomposition

If the compost pile isn't breaking down as quickly as expected, it may be too dry or lacking in microbial activity.

- Solution: Add more moisture if the pile is too dry, or mix in more nitrogen-rich materials (such as kitchen scraps) to boost microbial activity.

C. Pest Problems

Pests such as rodents or insects may be attracted to the compost pile if it contains food scraps or isn't properly sealed.

- Solution: Ensure that the compost pile is well-covered and avoid adding food waste to humanure compost piles. Use a secure, pest-resistant composting bin if needed.

16. Waste Minimization Strategies, Reduce, Reuse, and Recycle

Reducing waste is one of the simplest ways to minimize your environmental impact while living off the grid. By adopting sustainable practices like reusing materials and recycling effectively, you can drastically reduce the amount of waste your household produces.

1. The Importance of Waste Minimization in Off-Grid Living

Living off-grid often means limited access to waste management services like trash collection or recycling centers. Therefore, it's essential to create less waste from the start, reuse materials whenever possible, and properly dispose of or recycle whatever remains. Here's why waste minimization is key:

- Resource Conservation: The less waste you produce, the fewer raw materials and energy are needed to replace or process new goods.

- Reduced Environmental Impact: Less waste means fewer trips to the landfill and less pollution in the environment. Recycling helps conserve natural resources and reduce energy consumption.

- Lower Costs: Minimizing waste also means saving money by reusing materials and avoiding unnecessary purchases.

2. Reducing Waste at the Source

The best way to minimize waste is by reducing it at the source. This means being mindful of what you consume and taking steps to cut down on the materials you bring into your home.

A. Buying in Bulk and Avoiding Packaging

One of the main contributors to household waste is excessive packaging. By buying in bulk, you reduce the amount of packaging waste generated from individual items.

- Practical Tips:
    - Purchase dry goods like grains, beans, and pasta in large quantities to reduce packaging waste.
    - Store bulk items in reusable containers to avoid using plastic bags or wrappers.

B. Choosing Durable and Multi-Use Products

Invest in durable products that can be reused or repurposed rather than disposable items.

- Examples:
    - Replace single-use items (like paper towels) with washable cloth alternatives.
    - Choose stainless steel or glass containers instead of plastic, which wears down quickly.

C. Limiting Single-Use Plastics

Single-use plastics like bags, bottles, and utensils are major contributors to landfill waste and environmental pollution.

- Practical Tips:
    - Carry reusable bags, water bottles, and utensils when you go out.
    - Avoid buying products with excessive plastic packaging.

3. Reusing and Repurposing Materials

After reducing waste at the source, the next step is reusing or repurposing items that would otherwise be discarded. By giving materials a second life, you can reduce your need to buy new items and minimize what ends up in the trash.

A. Reusing Household Items

Before throwing something away, consider whether it can be reused for a different purpose.

- Examples:
    - Glass jars can be used for food storage or as planters.
    - Old clothes can be repurposed into rags or craft projects.
    - Wooden pallets can be transformed into furniture, garden beds, or shelving.

B. Repairing Instead of Replacing

In a consumer-driven society, people often throw away broken items rather than repairing them. Off-grid living encourages a different mindset—one that prioritizes repairing and maintaining goods.

- Practical Tips:

- Learn basic repair skills for common household items like clothing, furniture, and electronics.

- Keep tools and materials on hand for quick fixes, such as sewing kits, adhesives, and spare parts.

C. Creative Upcycling

Upcycling involves turning waste or discarded materials into something of higher value. This approach not only reduces waste but also adds creativity and functionality to your home.

- Examples:
  - Old metal cans can be turned into storage containers, plant holders, or lanterns.
  - Wooden crates can be transformed into bookshelves or side tables.

4. Recycling: Closing the Loop

For items that can't be reused or repurposed, recycling is the next best option. While recycling may not always be convenient in off-grid areas, it's important to develop a system that allows you to handle recyclable materials responsibly.

A. Setting Up a Home Recycling Station

A well-organized recycling station makes it easier to manage materials like paper, plastic, glass, and metals.

- Practical Tips:
  - Designate separate bins for each type of recyclable material.
  - Make sure items are clean and dry before placing them in the recycling bins to avoid contamination.

B. Recycling Organic Waste

Organic waste, such as food scraps and yard trimmings, can be composted rather than thrown away. This not only reduces the amount of waste you send to the landfill but also creates valuable compost for your garden.

- Composting: Set up a compost bin for kitchen scraps, leaves, and grass clippings. Regularly turn the compost to aerate it and speed up the decomposition process.

- Vermicomposting: For smaller spaces, consider vermicomposting—using worms to break down organic waste into rich compost.

C. Recycling Electronics

Electronics, or e-waste, contain valuable metals and other materials that can be recovered and reused. However, they also contain hazardous materials that need to be handled properly.

- Practical Tips:
  - If you're upgrading or replacing electronics, consider donating them to others who may still find them useful.
  - Look for e-waste recycling programs in your area, or send old electronics to certified e-waste recyclers.

17. Biogas Systems, Turning Organic Waste into Renewable Energy

Biogas systems are a sustainable way to turn organic waste into valuable energy. By converting waste like food scraps, animal manure, and plant material into biogas, you can generate a renewable source of fuel for cooking, heating, or even powering a generator. This not only reduces the amount of waste you send to landfills but also provides an off-grid energy source that complements solar and wind power.

1. What Is Biogas?

Biogas is a mixture of methane, carbon dioxide, and other gases produced when organic matter decomposes in an oxygen-free (anaerobic) environment. Biogas can be burned to produce heat or electricity and is often used in rural and off-grid communities as a sustainable energy source.

- How It Works: Organic waste is placed in an airtight digester, where bacteria break down the material and release biogas. The gas is captured and stored for later use, while the remaining solid waste can be used as fertilizer.

2. Setting Up a Biogas System

Setting up a biogas system requires careful planning and the right materials to ensure efficiency and safety. Here are the key steps to get started:

A. Choosing a Biogas Digester

The digester is the heart of the biogas system, where organic waste is broken down to produce gas. There are several types of digesters to choose from, depending on your needs and available resources.

- Fixed-Dome Digesters: These are underground systems with a fixed dome where gas is stored. They are sturdy, long-lasting, and suitable for larger households.

- Floating-Drum Digesters: These systems have a floating lid that rises as gas is produced. They are easier to maintain but less durable over time.

- Bag Digesters: Bag digesters use a flexible, sealed plastic bag to hold the waste and capture the biogas. They are ideal for smaller households or temporary setups.

B. Sourcing Organic Materials

Biogas systems rely on organic materials like food scraps, animal manure, and agricultural waste. It's important to have a steady supply of organic material to keep the system running efficiently.

- Best Sources of Organic Material:
  - Food waste (vegetable peelings, fruit scraps, leftover food)
  - Animal manure (from livestock like cows, chickens, or pigs)
  - Agricultural waste (crop residues, grass clippings)

C. Installing and Operating the System

Once the digester is in place and organic materials are available, the system can be set up to generate biogas.

- How to Operate: Add organic waste to the digester regularly, keeping the environment airtight. As the waste decomposes, biogas is produced and stored in the digester for later use. The system should be monitored for leaks or blockages, and the remaining solid waste (called digestate) should be removed periodically to use as fertilizer.

3. Benefits of Biogas Systems for Off-Grid Living

Biogas systems provide several benefits that make them ideal for off-grid homes looking for sustainable energy solutions.

A. Renewable Energy Source

Biogas is a renewable energy source that uses waste materials already present in your home or on your property. It reduces reliance on fossil fuels and complements other renewable energy sources like solar and wind.

B. Waste Reduction

By converting organic waste into energy, biogas systems help reduce the amount of waste going to landfills, lowering your environmental footprint.

C. Fertilizer Production

The byproduct of biogas production—digestate—is a nutrient-rich organic material that can be used as fertilizer for gardens, improving soil health and reducing the need for chemical fertilizers.

4. Using Biogas in the Home

Biogas can be used for several household applications, making it a versatile energy source.

A. Cooking

Biogas is often used for cooking, providing a clean and efficient fuel source that burns with minimal smoke or odor. It's a popular alternative to wood stoves or propane burners in off-grid homes.

B. Heating

Biogas can be used to fuel space heaters or water heaters, providing warmth and hot water during colder months. It's particularly useful in areas with limited access to firewood or other heating fuels.

C. Electricity Generation

In some cases, biogas can be used to generate electricity by powering a small generator. This can provide backup power during cloudy or windless days when solar or wind power is insufficient.

18. Hazardous Waste Management, Safe Disposal of Batteries, Chemicals, and Electronics

While minimizing and reusing waste is essential in off-grid living, certain materials—such as batteries, chemicals, and electronics—require careful handling and disposal to avoid environmental harm. Proper hazardous waste management ensures that these materials don't leach into the soil or water supply, protecting both your health and the environment.

1. Types of Hazardous Waste in Off-Grid Homes

Off-grid living often involves the use of materials that are considered hazardous waste when they are no longer useful. Here are some common examples:

- Batteries: Used batteries (including lead-acid, lithium-ion, and alkaline) can leak harmful chemicals into the environment if not properly disposed of.

- Chemicals: Household chemicals like pesticides, paints, and cleaning products often contain toxic substances that need careful handling.

- Electronics: E-waste, including old phones, computers, and appliances, contains metals like lead, mercury, and cadmium, which can pollute the environment if not disposed of properly.

2. Handling and Storing Hazardous Materials

To minimize the risks associated with hazardous waste, it's important to handle and store these materials properly.

A. Safe Storage of Batteries

Batteries should be kept in a cool, dry location, away from direct sunlight, moisture, or heat sources. For off-grid systems, regularly inspect battery banks to check for leaks or corrosion.

B. Handling Household Chemicals

Keep hazardous chemicals in their original containers with labels intact. Store them in a secure area, out of reach of children or pets, and away from food or water sources.

3. Disposal and Recycling of Hazardous Waste

Proper disposal of hazardous materials is critical to prevent environmental contamination.

A. Recycling Batteries

Many categories of batteries, including lead-acid and lithium-ion batteries used in off-grid solar systems, can be recycled. Contact local recycling centers or hazardous waste facilities to find out where you can safely dispose of used batteries.

B. Disposing of Chemicals

Check local regulations for the disposal of household chemicals. Some areas offer hazardous waste collection days, where residents can bring in chemicals for safe disposal.

C. Recycling Electronics

Many electronic devices can be refurbished or recycled to recover valuable materials like copper, gold, and silver. Find certified e-waste recyclers to handle old electronics safely.

Module E | No Grid Cooking, Heating, Cooling & Lighting Systems

19. Off-Grid Cooking Systems, Sustainable and Efficient Ways to Cook Without Power

Cooking off the grid requires a shift from conventional appliances like electric stoves and microwaves to more sustainable methods that harness natural energy sources. Whether you're using the sun's power or a fuel-efficient wood stove, off-grid cooking can be both resourceful and eco-friendly. This section explores three primary cooking systems—solar ovens, rocket stoves, and traditional wood-fired stoves—offering practical solutions to help you prepare meals without relying on electricity or gas.

Solar Ovens: Harnessing the Power of the Sun

One of the most sustainable ways to cook off-grid is using a solar oven, which harnesses the sun's energy to heat and cook food. Solar ovens are especially useful in sunny climates and can bake, roast, or dehydrate food without any fuel.

1. How Solar Ovens Work

Solar ovens rely on the greenhouse effect, using reflective surfaces to concentrate sunlight into a cooking chamber. The sunlight is converted into heat, which gets trapped inside the insulated chamber, allowing temperatures to rise and cook food slowly over time.

Key Components:

- Reflectors: Flat or curved surfaces that focus sunlight onto the cooking chamber.

- Cooking Chamber: An insulated box or container that traps heat.

- Transparent Cover: A glass or plastic cover that allows sunlight in while preventing heat from escaping.

Practical Tip: Solar ovens work best in areas with consistent sunlight. They are slower than traditional stoves, making them ideal for slow-cooking dishes like stews, casseroles, or baked goods.

2. Types of Solar Ovens

There are several types of solar ovens, each with different designs and capabilities. Here are the most common models you can use in an off-grid setting:

A. Box Solar Ovens

A box solar oven is a simple, insulated box with a glass cover and reflective panels that direct sunlight into the cooking chamber.

- Best For: Baking bread, roasting vegetables, and slow-cooking stews.

- Advantages: Easy to build or purchase, works well for low-heat, long-duration cooking.

- Limitations: Can take longer to reach high temperatures, especially in cloudy or cool conditions.

Example: The All-American Sun Oven is a popular box-style solar oven that can reach temperatures up to 400°F (200°C) and bake bread or cook stews with ease.

B. Parabolic Solar Cookers

A parabolic solar cooker uses a curved reflective surface to focus sunlight onto a central cooking pot, achieving higher temperatures faster than a box oven.
- Best For: Stir-frying, grilling, and boiling water quickly.
- Advantages: Can reach higher temperatures, making it suitable for more diverse cooking methods.
- Limitations: Requires frequent adjustments to stay aligned with the sun, making it less hands-off.

Example: The SolSource Parabolic Cooker can boil water in just minutes and grill food at high temperatures using only sunlight.

C. Panel Solar Cookers

Panel cookers are a combination of box and parabolic designs, using flat reflective panels to focus sunlight on a pot or pan.
- Best For: Low-heat cooking such as simmering or slow-cooking.
- Advantages: Lightweight, portable, and easy to use. Often more affordable than other solar oven types.
- Limitations: May struggle to achieve high temperatures, especially in less-than-ideal weather conditions.

3. Setting Up and Using a Solar Oven

Solar ovens are simple to set up and use, but there are a few key considerations to ensure you get the best performance.

Step 1: Choose the Right Location

Place your solar oven in a sunny spot with direct sunlight for several hours of the day. Ideally, the sun should hit the oven between late morning and mid-afternoon, when it's at its strongest.
- Practical Tip: Rotate your solar oven periodically to track the movement of the sun and maintain optimal cooking temperatures.

Step 2: Preheat the Oven

Just like a conventional oven, solar ovens need time to preheat. Depending on the weather, this can take anywhere from 20 minutes to an hour. Keep the oven closed during preheating to trap as much heat as possible.

Step 3: Monitor and Adjust as Needed

Solar ovens cook food more slowly than traditional stoves, and cooking times can vary based on sunlight intensity. Check on your food periodically, and adjust the angle of the reflectors to keep the oven aligned with the sun.

Practical Tip: Use dark, thin cookware in your solar oven to absorb and retain heat more effectively than light-colored or thick pots.

Rocket Stoves: Efficient Wood-Fired Cooking

Rocket stoves are a highly efficient way to cook with wood, making them ideal for off-grid living. Unlike traditional open fires, rocket stoves are designed to burn small amounts of wood cleanly and with minimal smoke, providing a faster and more efficient cooking method.

1. How Rocket Stoves Work

Rocket stoves use a small combustion chamber where wood or biomass is burned efficiently, drawing air through the bottom and up through the stove's chimney. The insulated design focuses heat directly on the cooking surface, requiring less fuel and reducing emissions.

Key Components:

- Combustion Chamber: Where the wood burns and heat is generated.

- Chimney: Helps draw in air and channel the heat upwards to the cooking surface.

- Fuel Feed: A small opening where wood is inserted into the combustion chamber.

Practical Tip: Rocket stoves are perfect for areas where wood is readily available. They require far less wood than an open fire, making them efficient for both cooking and heating.

2. Types of Rocket Stoves

There are several variations of rocket stoves, from simple DIY models to more advanced, commercially available designs.

A. DIY Rocket Stoves

Rocket stoves are relatively easy to build using materials like metal cans, bricks, or steel pipes. A simple DIY rocket stove can be constructed in a few hours with minimal tools.

- Best For: Budget-friendly, easy-to-build solutions for off-grid cooking.

- Advantages: Inexpensive and customizable, making them perfect for experimenting with different designs.

- Limitations: May not be as durable or efficient as professionally manufactured stoves.

Example: A basic rocket stove can be built using a metal can for the combustion chamber and an L-shaped pipe to direct the airflow.

B. Commercial Rocket Stoves

Commercial rocket stoves are designed for maximum efficiency and durability, often made from steel or cast iron. They come in a variety of sizes and designs, some of which include insulation and built-in cooking surfaces.

- Best For: Reliable and efficient cooking for off-grid homes or campsites.

- Advantages: Professionally built, offering superior efficiency and durability compared to DIY models.

- Limitations: More expensive than DIY stoves but often more efficient in the long run.

Example: The EcoZoom Versa Rocket Stove is a popular commercial model known for its efficiency, minimal smoke output, and ability to cook with small amounts of wood or biomass.

3. Cooking with a Rocket Stove

Rocket stoves heat up quickly and are perfect for tasks like boiling water, frying, or cooking meals in pots and pans.

Step 1: Gather Fuel

Rocket stoves are designed to burn small pieces of wood, twigs, or biomass. Gather dry, lightweight materials that will ignite easily and burn cleanly.

- Practical Tip: Use small, dry sticks or wood scraps for the most efficient burn. Avoid using wet or green wood, which produces more smoke and less heat.

Step 2: Light the Fire

Place a small amount of kindling in the combustion chamber and light it. Once the fire is established, feed small pieces of wood into the fuel feed, keeping the combustion chamber about half full for optimal airflow.

- Practical Tip: Use a firestarter like wax-coated paper or dry bark to help get the fire going quickly.

Step 3: Cook Your Meal

Once the stove is burning steadily, place your pot or pan on top of the stove. Rocket stoves generate intense heat quickly, so they're great for boiling water or stir-frying meals. Adjust the amount of fuel as needed to maintain the desired temperature.

Traditional Wood-Fired Stoves: Versatile Off-Grid Cooking

Wood-fired stoves have been used for centuries to cook food and heat homes. In an off-grid setting, a traditional wood-fired stove can provide reliable heat for cooking, especially in colder climates where warmth is also a priority.

1. The Versatility of Wood-Fired Stoves

Wood stoves are versatile in that they can both cook food and heat your living space. Many models come with flat surfaces for cooking or even built-in ovens for baking.

- Key Features:

- Cooking Surface: The flat top of the stove can be used to heat pots and pans.

- Oven: Some wood stoves include an oven compartment for baking.

- Heat Output: In addition to cooking, wood stoves are excellent for heating your home, especially in winter.

Practical Tip: Wood-fired stoves are particularly useful in colder climates where you need to cook and heat your home simultaneously.

2. Types of Wood-Fired Stoves

There are many different types of wood-fired stoves, from compact models designed for cabins to larger stoves that can heat an entire home while also cooking meals.

A. Compact Cooking Stoves

Compact wood-fired cooking stoves are smaller and designed primarily for cooking, making them ideal for tiny homes or outdoor kitchens.

- Best For: Small homes, tiny houses, or outdoor kitchens where space is limited.

- Advantages: Efficient for cooking without taking up too much space.

- Limitations: May not provide enough heat to warm an entire house.

Example: The Gstove Camping Stove is a compact, portable option that can be used for both heating and cooking in small off-grid spaces.

B. Multi-Purpose Wood Stoves

Larger wood-fired stoves are designed to serve both as a heating source and a cooking appliance. These stoves often include oven compartments and multiple cooking surfaces.

- Best For: Off-grid homes in cold climates where heating and cooking are both necessary.

- Advantages: Can heat an entire house while also providing a reliable cooking surface.

- Limitations: Larger models require more space and may not be suitable for smaller homes or warm climates.

Example: The Wood Cook Stove by Kitchen Queen is a robust option for off-grid homes, offering a spacious cooking surface, built-in oven, and enough heat to warm large living areas.

3. Cooking with a Wood-Fired Stove

Cooking on a wood-fired stove is straightforward, but it requires a bit of skill in managing the fire to achieve the right cooking temperature.

Step 1: Build a Fire

Start by building a fire in the stove's firebox. Use dry wood and kindling to establish a hot, steady flame. Adjust the air vents to control airflow and maintain a steady burn.

- Practical Tip: Split your wood into smaller pieces for faster, more efficient burning. Use larger logs for longer-lasting heat.

Step 2: Cook on the Surface or in the Oven

Once the stove is hot, you can place your pots and pans directly on the cooking surface or use the built-in oven to bake bread, casseroles, or roast vegetables.

- Practical Tip: The temperature on a wood stove surface varies depending on where the pot is placed. For high-heat cooking, place your pot directly over the firebox. For simmering, move it to the edge of the surface.

Step 3: Maintain the Fire

Keep the fire going as needed by adding wood periodically. For long-cooking dishes like stews or soups, use slow-burning hardwood logs to maintain consistent heat over time.

Conclusion: Choosing the Right Off-Grid Cooking System

Whether you're harnessing solar energy, using a fuel-efficient rocket stove, or relying on a traditional wood-fired stove, there are plenty of options for cooking off the grid. Each system offers unique advantages depending on your climate, resources, and cooking needs. By understanding how these systems work and how to use them effectively, you'll be well-equipped to prepare delicious, sustainable meals without relying on conventional power sources.

20. Off-Grid Heating Systems, Efficient and Sustainable Ways to Warm Your Home

Heating is a critical aspect of off-grid living, especially in colder climates where maintaining warmth without relying on conventional energy sources is essential. From efficient wood stoves to passive solar heating and thermal mass systems, there are several methods available to heat your off-grid home sustainably. In this section, we'll explore these options in detail, focusing on how to maximize efficiency while minimizing energy consumption.

Wood Stoves: A Classic and Reliable Heating Method

For centuries, wood stoves have been a primary source of heat in off-grid and rural homes. Efficient wood stoves are designed to burn wood cleanly and effectively, providing a steady source of warmth while reducing fuel consumption and emissions. With the right stove and setup, you can heat your home efficiently throughout the winter months.

1. How Efficient Wood Stoves Work

Modern wood stoves are designed to burn wood more efficiently than traditional open fireplaces, thanks to improved airflow and insulated chambers that maximize heat output while minimizing smoke and emissions. These stoves provide steady, radiant heat, making them an excellent option for off-grid homes.

Key Components:

- Firebox: Where wood is burned, generating heat.
- Baffle System: Helps to circulate hot air within the stove, increasing efficiency.
- Air Control: Allows you to adjust the airflow to regulate the intensity of the fire.
- Chimney: Vents smoke and exhaust gases safely outside the home.

Practical Tip: To get the most out of your wood stove, use seasoned hardwood, such as oak or maple, which burns hotter and longer than softwood.

2. Types of Efficient Wood Stoves

There are several types of wood stoves available, each designed to offer maximum efficiency and heat output while minimizing environmental impact.

A. Catalytic Wood Stoves

Catalytic wood stoves use a catalytic combustor to re-burn smoke and exhaust gases, resulting in a cleaner and more efficient burn. This process allows the stove to generate more heat from the same amount of wood.

- Best For: Off-grid homes where long-lasting heat is needed.
- Advantages: Catalytic stoves can maintain a steady heat output for longer periods, reducing the need to frequently add wood.
- Limitations: These stoves require regular maintenance to keep the catalytic combustor functioning properly.

Example: The Blaze King King 40 is a well-known catalytic wood stove that can burn for up to 40 hours on a single load of wood, making it ideal for long, cold winters.

B. Non-Catalytic Wood Stoves

Non-catalytic stoves don't use a combustor but instead rely on secondary combustion and an insulated firebox to achieve high efficiency. These stoves are simpler to maintain and are still highly effective at heating small to medium-sized homes.

- Best For: Those looking for a low-maintenance heating solution.
- Advantages: Less maintenance is required than catalytic models, and they still offer strong heat output.
- Limitations: They don't burn as efficiently as catalytic stoves, meaning you may need to add wood more frequently.

Example: The Jøtul F500 Oslo is a popular non-catalytic wood stove that combines efficiency with classic design, providing a reliable heat source for off-grid homes.

C. Pellet Stoves

Pellet stoves burn compressed wood or biomass pellets, offering a highly efficient and automated heating solution. These stoves are equipped with hoppers that feed pellets into the firebox automatically, reducing the need for constant refueling.

- Best For: Homes with access to pellet fuel and those seeking a more automated heating system.

- Advantages: Pellet stoves are highly efficient, produce minimal smoke, and require less frequent refueling than traditional wood stoves.

- Limitations: They rely on a small amount of electricity to operate the hopper and fans, which can be a limitation in fully off-grid settings without reliable backup power.

Example: The Harman P68 Pellet Stove is known for its efficiency and ease of use, capable of heating large spaces with minimal fuel.

3. Installing and Maintaining a Wood Stove

Proper installation and maintenance are key to ensuring that your wood stove operates safely and efficiently.

A. Stove Placement and Safety

When installing a wood stove, place it in a central location where the heat can radiate evenly throughout the home. Ensure there's enough clearance around the stove to prevent accidental fires, and use a heat-resistant floor pad under the stove if it's placed on combustible flooring.

- Ventilation: Install the chimney properly to vent exhaust gases safely. Make sure it's tall enough to create a strong draft and prevent smoke from backing up into your home.

Practical Tip: Consider using a heat-powered stove fan to circulate warm air throughout the room, especially in larger spaces.

B. Maintaining Your Wood Stove

Regular maintenance keeps your wood stove running efficiently and safely. Clean the stove and chimney regularly to remove creosote buildup, which can cause dangerous chimney fires.

- Creosote Removal: Use a chimney brush to clean the chimney at least once a year, or hire a professional to do it. Creosote buildup not only reduces efficiency but also poses a fire hazard.

- Inspecting Seals and Gaskets: Check the stove door gaskets for wear and replace them as needed to maintain an airtight seal, ensuring the fire burns efficiently.

Passive Solar Heating: Utilizing the Sun for Free, Clean Energy

Passive solar heating is a sustainable and cost-effective way to heat your home without using fuel or electricity. By designing your home to capture and store the sun's warmth, you can sustain a comfortable indoor climate year-round, especially during the colder months.

1. How Passive Solar Heating Works

Passive solar heating systems work by using the sun's energy to heat your home directly. This method relies on building design, materials, and orientation to maximize sunlight exposure and store heat.

Key Components:

- South-Facing Windows: These windows allow sunlight to enter and warm the interior of your home.

- Thermal Mass: Materials like concrete, stone, or brick absorb and store heat during the day, releasing it slowly at night.

- Insulation: Proper insulation helps retain the heat gained from the sun, reducing the need for additional heating sources.

Practical Tip: Homes in the Northern Hemisphere should maximize south-facing windows to capture the most sunlight during the day. In the Southern Hemisphere, north-facing windows should be prioritized.

2. Design Principles for Passive Solar Heating

To make the most of passive solar heating, your home should be designed with several key principles in mind.

A. Orientation and Window Placement

The orientation of your home plays a critical role in passive solar heating. Ideally, large windows should be placed on the south side of the house to capture the most sunlight. East- and west-facing windows can contribute, but they're less effective and can cause unwanted heat gain during the summer.

- Window Size: Larger windows maximize solar gain, but they should be combined with thermal mass to prevent overheating during the day.

B. Thermal Mass

Thermal mass refers to materials that absorb and store heat. When sunlight enters your home, it warms the thermal mass, which then releases heat slowly throughout the day and night.

- Best Materials: Concrete floors, brick walls, and stone surfaces are excellent examples of thermal mass. These materials are not only durable but also effective at storing heat.

- Placement: Place thermal mass near south-facing windows where it will receive direct sunlight. It's essential that the mass is thick enough to store substantial heat without causing temperature fluctuations.

C. Insulation and Air Sealing

To keep the heat gained from the sun, your home must be well-insulated. Insulation helps maintain a stable indoor temperature, preventing heat from escaping at night.

- Best Practices: Use high-quality insulation in walls, floors, and ceilings to reduce heat loss. Seal air leaks around windows, doors, and other openings to prevent cold air from entering.

3. Passive Solar Heating Strategies

There are several practical strategies you can use to incorporate passive solar heating into your off-grid home.

A. Sunspaces or Solar Rooms

A sunspace is a room or enclosed porch that captures sunlight and transfers it to the rest of the home. These spaces act as greenhouses, warming the air during the day and allowing heat to circulate into adjacent rooms.

- Practical Tip: Use thermal mass in your sunspace (such as concrete flooring or masonry walls) to store heat and release it slowly throughout the night.

B. Trombe Walls

A Trombe wall is a thick wall (made of concrete, stone, or brick) placed just behind a large south-facing window. The wall absorbs solar heat during the day and radiates it into the home at night.

- How It Works: Sunlight passes through the window and heats the wall. Over time, the wall releases heat into the living space, maintaining warmth even after the sun sets.

C. Shading and Overhangs

While passive solar heating is ideal for the colder months, it's important to prevent overheating in the summer. Shading devices like overhangs or retractable awnings help block sunlight during the hot months.

- Practical Tip: Design overhangs so that they block the high summer sun but allow the lower winter sun to enter your home, balancing heating and cooling needs throughout the year.

Thermal Mass Heating: Storing Heat for Long-Lasting Warmth

Thermal mass heating is a highly effective way to retain and distribute heat in off-grid homes. This method involves using materials like concrete, brick, or stone to absorb and store heat, which is then slowly released over time. When combined with wood stoves or passive solar heating, thermal mass systems can significantly reduce the need for additional heating.

1. The Science of Thermal Mass

Thermal mass works by absorbing heat during the day and releasing it gradually over several hours or even days. The denser the material, the more heat it can store. When used in combination with a heating source, thermal mass creates a stable, comfortable indoor climate.

- How It Works: During the day, thermal mass absorbs excess heat generated by a wood stove or captured through passive solar heating. As the temperature drops at night, the mass radiates this stored heat back into the room, reducing temperature fluctuations.

2. Integrating Thermal Mass into Your Home Design

Incorporating thermal mass into your off-grid home can be done in several ways, depending on your building materials and layout.

A. Masonry Heaters

Masonry heaters are large, efficient wood stoves built from materials like brick or stone. Unlike traditional wood stoves, masonry heaters are designed to absorb and store heat, releasing it gradually over time.

- Best For: Homes in cold climates that require long-lasting heat.

- Advantages: Once heated, a masonry heater can radiate warmth for 12 to 24 hours with a single fire.

Practical Tip: Masonry heaters require professional installation but provide one of the most efficient ways to heat an off-grid home, especially when combined with thermal mass features like stone walls or floors.

B. Concrete or Stone Floors

Concrete or stone floors are excellent choices for off-grid homes because they function as both structural elements and thermal mass. These materials absorb heat during the day and release it slowly at night.

- Best For: Homes with passive solar heating systems or wood stoves.

- Advantages: Concrete and stone are durable, low-maintenance materials that provide natural heat retention without the need for additional energy.

3. Combining Thermal Mass with Other Heating Methods

Thermal mass heating is most effective when combined with other heating methods, such as wood stoves or passive solar heating. By strategically placing thermal mass near heat sources, you can store excess heat and release it when needed, creating a more energy-efficient home.

- Practical Tip: Position thermal mass elements (like concrete walls or floors) near windows that receive direct sunlight or next to your wood stove to maximize heat absorption and distribution.

Conclusion: Efficient Heating Solutions for Off-Grid Homes

By using a combination of efficient wood stoves, passive solar heating, and thermal mass systems, you can keep your off-grid home warm and comfortable throughout the year without relying on conventional energy sources. Each method has its unique benefits, from the high efficiency of catalytic wood stoves to the cost-effectiveness of passive solar design. By choosing the right combination for your climate and home layout, you can create a heating system that's both sustainable and effective, ensuring a comfortable living environment in even the harshest weather conditions.

21. Natural Cooling Systems, Sustainable Ways to Keep Your Off-Grid Home Cool

Cooling an off-grid home requires energy-efficient strategies that make use of natural elements like shade, air circulation, and thermal mass. Unlike conventional air conditioning systems, natural cooling methods reduce or eliminate the need for electricity while maintaining comfort in hot climates. In this section, we'll explore practical approaches like earth tubes, shading, and effective insulation to keep your home cool and comfortable without relying on the grid.

Earth Tubes: Harnessing Subterranean Temperatures for Cooling

Earth tubes, also known as ground-coupled heat exchangers, are an innovative way to cool (and sometimes heat) an off-grid home by taking advantage of the stable temperatures found underground. By circulating air through underground pipes, earth tubes help cool indoor air during the summer by transferring heat to the cooler ground below.

1. How Earth Tubes Work

Earth tubes function by utilizing the consistent temperature of the earth several feet below the surface, which is typically cooler than the outdoor air in the summer. Air is drawn into the home through a series of underground pipes, where it cools before being released inside.

Key Components:

  - Inlet Pipe: Draws warm outside air into the system.

  - Underground Pipes: Buried several feet below the surface, where they transfer heat from the air to the surrounding soil.

  - Outlet: Delivers cooled air into the home.

  - Fan System (optional): In passive systems, natural airflow drives the process, but some setups use fans to increase air circulation.

Practical Tip: Earth tubes are most effective in dry climates with significant temperature differences between day and night. They also work best in well-insulated homes that retain cool air.

2. Benefits of Earth Tube Systems

Earth tubes are an energy-efficient way to cool your home, especially when compared to traditional air conditioning systems. Here's why they're a valuable option for off-grid living:

  - Energy Efficiency: Since earth tubes rely on natural temperature differences, they require minimal energy to operate. In passive systems, no electricity is needed at all.

  - Long-Term Sustainability: Once installed, earth tube systems have minimal maintenance costs and provide long-lasting cooling without ongoing energy consumption.

  - Air Quality: Earth tubes help circulate fresh air into your home, improving indoor air quality while cooling the space.

3. Installing an Earth Tube System

Setting up an earth tube system requires careful planning and design to ensure optimal performance. Here are the key steps involved in installing an earth tube system in your off-grid home.

A. Designing the System

The design of an earth tube system depends on factors like your local climate, soil type, and the size of your home. The length, depth, and diameter of the pipes will all affect how well the system works.

- Pipe Depth: For most climates, pipes should be buried at least 6 to 10 feet below the surface to take advantage of the earth's stable temperature. In very hot climates, deeper pipes may be required.

- Pipe Material: Use materials like PVC or polyethylene that are resistant to corrosion and can withstand the weight of the soil above.

- Pipe Slope: Install pipes with a slight downward slope to allow condensation to drain out of the system and prevent moisture buildup.

B. Installing the System

Once the design is finalized, the pipes are laid underground, and the inlet and outlet points are connected to your home.

- Excavation: Dig trenches for the pipes, making sure to reach the appropriate depth for your climate.

- Pipe Installation: Lay the pipes in the trench, ensuring they are properly sealed to prevent leaks. Backfill the trench and compact the soil around the pipes.

- Airflow Control: Install any necessary fans or valves to regulate airflow and prevent backdrafts.

Shading: A Simple and Effective Cooling Strategy

One of the easiest and most cost-effective ways to cool your off-grid home is through shading. By blocking direct sunlight from entering your home, you can significantly reduce indoor temperatures, making it easier to keep your living space cool.

1. The Importance of Shading in Cooling

Sunlight streaming through windows can quickly raise the temperature inside your home. Shading prevents this heat gain by blocking the sun's rays before they reach your home's interior.

- How It Works: Shading elements like overhangs, awnings, and trees block sunlight during the hottest part of the day, keeping your home cooler. By reducing the need for active cooling methods, shading helps conserve energy and improve comfort.

2. Types of Shading Systems

There are several shading methods you can incorporate into your home design. These range from simple solutions like planting trees to more complex systems like retractable awnings or adjustable shutters.

A. Overhangs and Roof Extensions

A properly designed overhang or roof extension can block the high summer sun while still allowing the lower winter sun to enter your home. This passive solar strategy helps maintain a comfortable indoor temperature throughout the year.

- Best For: Homes with large, south-facing windows in warm climates.

- Design Tip: Calculate the angle of the sun during the summer months and design overhangs accordingly to provide optimal shading.

B. Retractable Awnings and Shutters

Retractable awnings and adjustable shutters give you control over how much sunlight enters your home. These systems can be adjusted based on the time of day and the season, allowing you to block the sun when needed and let in light during cooler hours.

- Best For: Homes with variable sun exposure or where flexibility is needed.

- Advantages: Retractable awnings allow you to enjoy shade when the sun is strong and retract them during cooler parts of the day, maximizing natural light without overheating.

Example: The SunSetter Retractable Awning is a popular option for homeowners looking for an adjustable shading solution that can be extended or retracted with ease.

C. Natural Shading with Trees and Plants

Trees and plants offer an eco-friendly and aesthetic way to shade your home naturally. Deciduous trees, which lose their leaves in winter, provide shade in the summer while allowing sunlight to reach your home during colder months.

- Best For: Homes in warm or temperate climates where tree growth is feasible.

- Advantages: Trees not only provide shade but also help cool the surrounding air through the process of transpiration.

Practical Tip: Plant trees on the south and west sides of your home to block the most intense sunlight during the hottest part of the day.

Effective Insulation: Keeping Cool Air In and Hot Air Out

While insulation is often associated with keeping homes warm, it's just as important for cooling. Effective insulation helps keep hot air out and cool air in, making your home more energy-efficient and comfortable during the summer months.

1. The Role of Insulation in Cooling

Insulation works by slowing the transfer of heat through walls, ceilings, and floors. In the summer, it prevents hot outdoor air from entering your home, reducing the need for active cooling systems.

Key Areas to Insulate:

- Walls: Insulate walls to block heat from the sun.

- Attics and Roofs: Attic insulation is particularly important for cooling, as hot air tends to accumulate in the upper parts of your home.

- Windows: Use double-glazed or Low-E windows to reduce heat gain while still allowing natural light to enter.

Practical Tip: In warm climates, opt for reflective insulation materials that deflect heat away from your home, such as foil-backed insulation.

2. Types of Insulation for Cooling

There are several types of insulation materials that are particularly effective for cooling. Choosing the right material depends on your climate, budget, and building design.

A. Reflective Insulation

Reflective insulation is designed to reflect heat rather than absorb it, making it ideal for keeping homes cool in hot climates. It typically consists of a reflective foil layer that blocks radiant heat.

- Best For: Homes in hot, sunny climates where reducing heat gain is a priority.

- Advantages: Reflective insulation is easy to install and highly effective at reducing heat transfer through roofs and attics.

B. Spray Foam Insulation

Spray foam insulation expands to fill gaps and cracks in walls and ceilings, creating an airtight barrier that prevents hot air from entering and cool air from escaping.

- Best For: Homes with hard-to-reach areas or irregular spaces that need complete coverage.

- Advantages: Spray foam provides both insulation and an air seal, making it highly effective for cooling and energy efficiency.

C. Cellulose Insulation

Cellulose insulation is made from recycled paper products and treated with fire retardants. It's an eco-friendly option that provides good thermal resistance and can be used in walls, attics, and floors.

- Best For: Homes looking for a sustainable insulation option.

- Advantages: Cellulose insulation is affordable and provides good thermal and acoustic performance, helping keep homes cool and quiet.

3. Insulation Installation Tips for Maximizing Cooling Efficiency

Proper installation is key to maximizing the cooling benefits of insulation. Here are some tips to ensure your insulation is working effectively.

A. Seal Air Leaks

Even the best insulation won't be effective if your home has air leaks. Check for gaps around windows, doors, and other openings, and seal them with caulk or weatherstripping to prevent hot air from infiltrating your home.

Practical Tip: Use expanding foam to seal larger gaps or cracks around pipes and electrical outlets.

B. Insulate Attics and Roofs

Heat tends to rise, so your attic and roof are often the hottest parts of your home during the summer. Properly insulating these areas can prevent heat from entering and lower indoor temperatures significantly.

Practical Tip: In addition to insulation, consider installing a radiant barrier in your attic to reflect heat away from your home.

Combining Cooling Methods for Maximum Efficiency

The most effective cooling strategies combine multiple approaches to create a comfortable indoor environment with minimal energy use. Here's how you can integrate these natural cooling methods into your off-grid home.

A. Earth Tubes and Insulation for Efficient Cooling

Using earth tubes to pre-cool incoming air, combined with high-quality insulation, helps maintain a cool indoor temperature even on the hottest days. The insulation will keep the cool air inside while preventing hot outdoor air from infiltrating your home.

B. Shading and Insulation for Passive Cooling

Combine shading techniques like overhangs or awnings with proper insulation to block direct sunlight and reduce heat gain. This minimizes the need for active cooling systems and keeps your home comfortable naturally.

Practical Tip: In climates with hot summers and cool winters, consider using retractable shading systems that can be adjusted seasonally for optimal performance.

C. Cross-Ventilation for Natural Cooling

Take advantage of natural air movement by incorporating cross-ventilation into your home design. By strategically placing windows and vents, you can create a natural breeze that helps cool your home without the need for fans or air conditioning.

Best Practices: Place windows on opposite sides of the house to create a pathway for air to flow through. Use ceiling or attic vents to allow hot air to escape while drawing cooler air in from the lower levels of the house.

Conclusion

By using a combination of natural cooling methods like earth tubes, shading, and effective insulation, you can keep your off-grid home comfortable during the hottest months without relying on energy-intensive air conditioning. These strategies not only lower energy usage but also contribute to a healthier, more comfortable living environment that works in harmony with the natural world. Whether you're building a new off-grid home or retrofitting an existing one, these cooling methods offer sustainable and efficient solutions for year-round comfort.

22. Off-Grid Lighting Systems, Illuminating Your Home Without Electricity

Lighting is an essential part of any home, and when living off the grid, traditional electric lighting may not always be feasible. However, there are many alternative lighting solutions available that don't rely on the power grid, including solar-powered lights, oil lamps, and handmade candles. In this section, we'll explore these options, offering practical advice on how to illuminate your off-grid home efficiently and sustainably.

Solar-Powered Lighting: Clean, Efficient, and Renewable

Solar-powered lighting is one of the most popular and accessible solutions for off-grid living. Harnessing the energy of the sun, solar lights provide clean and reliable illumination without any dependence on external electricity sources. Whether used for outdoor lighting, interior fixtures, or emergency lights, solar power offers a sustainable solution for long-term energy independence.

1. How Solar Lighting Systems Work

Solar lights operate by converting sunlight into electricity via solar panels. This electricity is stored in a rechargeable battery, which powers the lights during the evening and nighttime hours.

Key Components:

- Solar Panel: Captures sunlight and converts it into electricity.

- Rechargeable Battery: Stores the electricity generated by the solar panel for use when the sun isn't shining.

- LED Lights: Energy-efficient light bulbs that provide bright illumination while consuming minimal power.

Practical Tip: Solar lights are most effective when placed in locations with consistent sunlight. For optimal performance, ensure the solar panels are free of dirt and debris, and position them to face the sun for the majority of the day.

2. Types of Solar-Powered Lighting Systems

Solar lighting comes in a variety of designs to meet different needs, from small garden lights to full home lighting systems. Here are the most common types:

A. Stand-Alone Solar Lights

These are individual lights with built-in solar panels and batteries. Stand-alone solar lights are ideal for outdoor use, such as garden paths, driveways, or patios.

- Best For: Outdoor illumination in areas where wiring is difficult or impractical.

- Advantages: Easy to install, no need for wiring, and fully self-sufficient.

Example: The Ring Solar Pathlight is a durable and bright stand-alone option that automatically turns on at dusk and charges during the day.

B. Solar-Powered Indoor Lighting

For indoor use, solar panels can be installed on the roof or another sunny location, with wiring running to lights throughout the house. These systems are more complex but provide consistent lighting for multiple rooms.

- Best For: Whole-house lighting in off-grid homes where solar panels are already in place for other energy needs.
- Advantages: Can provide a steady and renewable source of light for an entire home.

Practical Tip: When setting up solar-powered indoor lighting, use LED bulbs wherever possible. They consume far less power than incandescent or fluorescent bulbs, allowing your solar energy system to last longer and light your home more efficiently.

C. Portable Solar Lights

Portable solar lights are versatile and can be moved around as needed. These are great for camping, emergency preparedness, or temporary lighting in off-grid homes.

- Best For: Flexible lighting needs, such as outdoor activities, power outages, or temporary workspaces.
- Advantages: Lightweight, easy to carry, and can be charged during the day for use at night.

Example: The LuminAID Solar Lantern is a foldable, portable lantern that charges in the sun and provides bright, long-lasting light, making it ideal for outdoor activities and emergencies.

3. Installing and Maintaining Solar Lighting Systems

Solar lighting systems are relatively easy to install, but there are a few important considerations to ensure they function properly.

A. Choosing the Right Location for Solar Panels

Solar panels need direct sunlight to charge efficiently. Position your solar panels in a spot that gets at least six hours of sunlight per day, and avoid placing them under trees, eaves, or other obstructions that can cast shadows.

Practical Tip: In climates with limited sun exposure, consider using a solar tracker, which adjusts the angle of the solar panels throughout the day to maximize energy absorption.

B. Cleaning and Maintaining Solar Panels

Over time, dust, dirt, and debris can accumulate on your solar panels, reducing their efficiency. Cleaning the panels every few months will ensure they're operating at peak performance.

- How to Clean: Use a soft cloth or sponge with water and mild soap to gently clean the surface of the solar panels. Avoid using harsh chemicals or abrasive materials, which can scratch or damage the panels.

Oil Lamps: A Traditional and Reliable Lighting Option

Oil lamps have been used for centuries as a reliable source of light, and they remain a practical option for off-grid living. These lamps burn fuel, such as kerosene, olive oil, or lamp oil, and provide a soft, warm glow that's ideal for indoor or outdoor use.

1. How Oil Lamps Work

Oil lamps consist of a reservoir that holds fuel and a wick that draws the fuel up to the flame. As the wick burns, the fuel is vaporized and combusted, producing light and heat.

Key Components:

- Wick. Draws fuel from the reservoir and provides a controlled burn.

- Fuel Reservoir: Holds the oil or kerosene used for combustion.

- Chimney: A glass or metal enclosure that helps direct the airflow and protects the flame from wind.

Practical Tip: Oil lamps are best suited for indoor use or sheltered outdoor spaces, as wind and rain can extinguish the flame. For maximum brightness, keep the wick trimmed to about 1/8 inch above the burner.

2. Types of Oil Lamps

There are several different types of oil lamps available, each suited for different lighting needs.

A. Traditional Kerosene Lamps

Kerosene lamps are one of the most common types of oil lamps and can provide long-lasting light with a small amount of fuel. These lamps are durable and ideal for emergency lighting or as a backup in off-grid homes.

- Best For: Emergency lighting or backup light sources.

- Advantages: Kerosene is widely available, and these lamps are inexpensive to operate.

- Limitations: Kerosene produces smoke and an odor, so proper ventilation is necessary.

B. Olive Oil Lamps

Olive oil lamps offer a cleaner, more sustainable option for oil-based lighting. Olive oil burns more cleanly than kerosene, producing little to no smoke or odor, making it a healthier option for indoor use.

- Best For: Indoor lighting where air quality is a concern.

- Advantages: Clean-burning and made from a renewable resource.

- Limitations: Olive oil burns at a lower temperature than kerosene, so these lamps may not be as bright.

C. Pressurized Oil Lamps

Pressurized oil lamps, such as Coleman lanterns, use pressurized fuel (typically kerosene or white gas) to create a bright, steady flame. These lamps are often used for outdoor activities like camping but can also provide reliable lighting for off-grid homes.

- Best For: Outdoor lighting or bright, long-lasting indoor lighting.

- Advantages: Provides much brighter light than traditional oil lamps.

- Limitations: Requires pressurized fuel, which can be more expensive and harder to find in some areas.

3. Safely Using and Maintaining Oil Lamps

While oil lamps are effective, they must be used safely to prevent accidents such as fires or fuel spills.

A. Choosing the Right Fuel

It's important to use the correct fuel for your lamp to ensure safe operation. Never use gasoline or other highly flammable liquids in oil lamps, as they can create a fire hazard.

- Recommended Fuels: Kerosene, olive oil, and specially designed lamp oils are the safest and most efficient fuels to use.

- Ventilation: Always use oil lamps in well-ventilated areas to avoid a buildup of carbon monoxide, especially when using kerosene.

B. Trimming the Wick

The wick in an oil lamp must be trimmed regularly to maintain a steady, even flame and prevent soot buildup. A well-trimmed wick will provide brighter light and use less fuel.

- Practical Tip: Keep the wick about 1/8 inch long for an optimal flame. If the flame flickers or emits smoke, trim the wick to a shorter length.

Candle-Making: An Affordable and DIY Lighting Solution

Handmade candles offer a simple, affordable, and creative way to provide light in an off-grid home. Whether made from beeswax, soy wax, or paraffin, candles are versatile and can be customized to suit your lighting needs.

1. How to Make Candles at Home

Making candles at home is a straightforward process that requires only a few basic materials. Here's a simple guide to DIY candle-making:

Materials Needed:

- Wax: Choose from beeswax, soy wax, or paraffin wax.

- Wicks: Cotton or hemp wicks are the most common.

- Molds or Containers: You can use anything from jars to silicone molds.

- Double Boiler: For melting the wax.

- Fragrance (optional): Essential oils or candle fragrances can be added for a pleasant scent.

Step 1: Melt the Wax

Use a double boiler to melt the wax slowly over low heat. Stir gently until the wax is fully melted.

Step 2: Prepare the Wick

While the wax is melting, secure the wick in the center of your mold or container. You can use a small dab of hot wax or a wick holder to keep the wick in place.

Step 3: Pour the Wax

Once the wax is fully melted, pour it into your container or mold, making sure to leave about half an inch of space at the top. Allow the candle to cool and harden completely.

Step 4: Trim the Wick

After the candle has hardened, trim the wick to about 1/4 inch to ensure a steady flame.

2. Types of Candle Wax

The type of wax you use for candle-making affects how long the candle will burn and the quality of the light it produces.

A. Beeswax

Beeswax candles burn cleanly and produce a warm, natural glow. They also emit a subtle honey scent, making them a great choice for indoor use.

- Best For: Long-burning, clean light with a natural scent.
- Advantages: Renewable, non-toxic, and produces little soot.
- Limitations: Beeswax is more expensive than other types of wax.

B. Soy Wax

Soy wax is a popular alternative to paraffin, made from natural soybean oil. Soy candles burn more slowly than paraffin candles and are available in a variety of shapes and sizes.

- Best For: Affordable and eco-friendly candles.
- Advantages: Renewable, biodegradable, and burns longer than paraffin.
- Limitations: Soy wax is softer, which means it may require containers or molds to hold its shape.

C. Paraffin Wax

Paraffin candles are the most widely available and affordable, but they do produce more soot than natural waxes. They are best used for short-term lighting needs.

- Best For: Inexpensive candles for short-term or emergency lighting.
- Advantages: Readily available and comes in a variety of colors and scents.
- Limitations: Produces soot and is derived from petroleum, making it less eco-friendly.

3. Storing and Using Handmade Candles

Handmade candles are easy to store and can provide emergency lighting when needed.

A. Safe Storage

Store candles in a cool, dry place away from direct sunlight to prevent them from melting or warping.

B. Using Candles Safely

Always use candles on a stable, heat-resistant surface and never leave them unattended. Keep them away from flammable materials and ensure proper ventilation when burning multiple candles indoors.

Practical Tip: Place candles inside lanterns or hurricane lamps to protect the flame from wind and reduce the risk of fire.

Module F | Fortification & Security

23. Fortification, Securing Your Off-Grid Home Against External Threats

Fortifying your home is a critical component of living off the grid, where self-reliance often means ensuring your safety without access to immediate external help. Whether you're preparing for potential intruders or natural disasters, a fortified home can provide peace of mind and a strong defensive position. This section will cover various strategies for assessing and addressing vulnerabilities in your home, reinforcing weak points, and setting up comprehensive security measures to protect your off-grid sanctuary.

Assessing Home Vulnerabilities: Identifying Weak Points

Before you begin fortifying your home, it's essential to identify potential vulnerabilities. Knowing where your home is most at risk allows you to prioritize improvements and address critical weaknesses.

1. Conducting a Security Audit

A security audit involves a thorough inspection of your home's physical layout, materials, and entry points to determine areas that could be exploited by intruders or affected by natural disasters.

- Walkthrough: Start by walking through your home and property, taking note of potential weak spots like unreinforced doors, single-pane windows, and easily accessible entry points (such as basement windows or skylights).

- Natural Barriers: Assess the natural landscape around your home—hills, trees, or waterways—that could either provide cover for intruders or be used to your advantage.

- Accessibility: Evaluate how easily someone could access your home or land. Look at driveways, paths, and any hidden entrances that could be exploited.

Practical Tip: Perform this audit from the perspective of a potential intruder. Look for areas that seem easy to access and imagine how you would break into your own home.

2. Entry Points: Doors, Windows, and More

Most intruders attempt to enter homes through easily accessible entry points, such as doors, windows, and garages. Reinforcing these areas can significantly reduce your home's vulnerability.

- Doors: Weak doors, especially those with hollow cores, are easy to break down. Prioritize strengthening your doors by installing solid wood or steel-core options, reinforcing frames, and adding deadbolt locks.

- Windows: Single-pane windows are easy to shatter. Consider upgrading to double-pane, tempered glass, or adding security film to reinforce windows without sacrificing visibility.

- Garage Doors: These can be a weak point, especially in homes with attached garages. Install heavy-duty locks and reinforce the garage door with bracing or additional locks.

Fortification Strategies: Strengthening Your Home's Defenses

Once vulnerabilities have been identified, the next step is to strengthen your home's defenses. From reinforcing walls to securing entry points, fortification strategies will help you create a resilient barrier against threats.

1. Reinforcing Doors and Windows

Doors and windows are the most common entry points for intruders. Here's how to make them as secure as possible.

A. Reinforced Doors

- Solid Core or Steel Doors: Replace weak exterior doors with solid wood or steel-core doors to prevent them from being kicked in or easily broken.

- Deadbolt Locks: Install heavy-duty deadbolt locks with long screws that extend into the door frame. This will make it much harder for intruders to force the door open.

- Door Frames: Reinforce door frames with metal plates or long screws to prevent them from splitting under pressure. Consider installing security hinges that can't be easily removed.

B. Reinforced Windows

- Security Film: Apply security film to windows to prevent them from shattering easily. Even if the glass breaks, the film will hold the fragments together, making it more difficult for intruders to gain entry.

- Window Bars or Grills: For high-risk areas, installing bars or grills can provide an extra layer of protection. Ensure these are securely attached and can't be easily removed.

- Window Locks: Upgrade window locks to more secure versions. Consider adding secondary locks or pins that prevent windows from being opened from the outside.

Practical Tip: Consider using laminated or polycarbonate glass for particularly vulnerable windows, as these materials are resistant to impact and much harder to break.

2. Securing Walls and Roofs

Reinforcing the structural elements of your home can prevent intruders from breaking through walls or entering through less obvious points of entry like the roof.

- Exterior Walls: Reinforce exterior walls with stronger materials like concrete, stone, or reinforced bricks. These materials are not only fire-resistant but also offer excellent protection against forced entry.

- Roof Access: If your home has an accessible roof, make sure it is secure. Consider adding spikes, mesh, or barbed wire to discourage entry. Ensure that skylights are reinforced with security film or barred from inside.

Fencing and Gates: Setting Up Perimeter Security

A strong, well-designed perimeter fence or gate can deter intruders before they even reach your home. Fencing also provides privacy, marking your territory and making it clear that entry is restricted.

1. Choosing the Right Fencing Material

The type of fencing you choose depends on your security needs, budget, and local regulations. Here are a few options:

- Chain-Link Fencing: A cost-effective option, chain-link fences are durable and difficult to climb. You can enhance security by adding barbed wire or spikes at the top.

- Wooden Fencing: Solid wooden fences provide more privacy and can be reinforced with steel supports or backed with chain-link to make them more difficult to break.

- Metal or Iron Fences: For higher security, metal fences—such as wrought iron—offer both durability and an imposing appearance. These fences are difficult to climb or cut through.

2. Gates and Entrances

Gates are often the first line of defense for your home, so it's essential that they are secure.

- Automatic Gates: For added security, install automatic gates that can be opened remotely, minimizing the risk of being exposed while opening or closing the gate manually.

- Reinforced Gates: If you prefer manual gates, ensure they are made from strong materials like steel and equipped with heavy-duty locks or latches. Add spikes or barbed wire at the top to prevent climbing.

Practical Tip: Consider adding motion sensors or alarms to your gate to alert you if someone tries to tamper with it.

Safe Rooms and Hiding Spots: Building Secure Shelter in Emergencies

Having a safe room or hiding spot can provide a critical refuge in the event of a home invasion or natural disaster. These spaces are designed to withstand forced entry and protect you while you wait for help or the threat to pass.

1. Building a Safe Room

A safe room is a fortified room within your home designed to provide maximum security. These rooms are reinforced with steel or concrete and equipped with secure doors and locks to prevent intruders from gaining access.

- Location: Safe rooms are typically located in basements, closets, or central areas of the home. Choose a space that is easy to access in an emergency but difficult for intruders to find.

- Reinforced Walls and Doors: Use steel or reinforced concrete to create impenetrable walls. Install a solid steel door with a heavy-duty lock or deadbolt.

- Emergency Supplies: Stock your safe room with essentials like food, water, a first-aid kit, communication devices (such as a battery-powered radio or satellite phone), and self-defense tools.

2. Discreet Hiding Spots

If you don't have the resources to build a full safe room, creating discreet hiding spots around your home can offer temporary protection in case of an emergency.

- Concealed Spaces: Convert spaces under stairs, behind bookshelves, or in attics into hiding spots. These areas should be difficult to find but easy for you to access quickly.

- Camouflaged Entrances: Use false walls, hidden doors, or furniture to disguise the entrance to your hiding spot. Ensure that it blends seamlessly with its surroundings to avoid detection.

DIY vs. Professional Fortification: Making the Right Choice

When it comes to fortifying your home, you'll need to decide whether to tackle the project yourself or hire professionals. Each approach has its advantages, depending on your budget, skills, and security needs.

1. DIY Fortification

For those who are handy or on a tight budget, DIY fortification can be a cost-effective solution. Many home security improvements, such as installing deadbolts or reinforcing windows, can be done with basic tools and materials.

- Advantages: Lower cost, complete control over the project, and the ability to customize fortifications to suit your specific needs.

- Limitations: Some fortifications, like structural reinforcements or safe rooms, may require advanced skills or specialized tools. In these cases, professional assistance is recommended.

Practical Tip: If you choose to go the DIY route, invest in high-quality materials and take the time to research best practices to ensure your home is properly secured.

2. Hiring Professionals

For larger projects or more complex security needs, hiring professionals ensures that the job is done safely and to a high standard. This is particularly important for structural fortifications, safe rooms, or advanced security systems.

- Advantages: Expertise in home security, high-quality materials, and adherence to local building codes and safety regulations.

- Limitations: Higher cost and less hands-on control over the project.

Practical Tip: When hiring professionals, get multiple quotes and check references to ensure you're working with reputable contractors who specialize in security fortifications.

Shelters: Underground Bunkers and Safe Spaces

For those concerned about severe threats—such as natural disasters, civil unrest, or even nuclear attacks—building a shelter offers the ultimate level of protection. These structures are designed to keep you safe in extreme circumstances.

1. Underground Bunkers

Underground bunkers provide maximum protection from a wide range of threats, including natural disasters, nuclear fallout, and armed intruders. These bunkers are built below ground, often reinforced with concrete or steel, and designed to be self-sufficient.

- Advantages: Complete isolation from external threats, including blasts, fires, and extreme weather conditions.

- Limitations: High construction costs and the need for specialized knowledge and materials.

Bunker Design Considerations

When designing a bunker, consider the following:

- Ventilation: Ensure that the bunker has a reliable air filtration system to provide fresh air and filter out contaminants.

- Water and Food Storage: Include enough water and non-perishable food to last for several weeks or months, depending on the anticipated threat.

- Energy Source: Install a renewable energy source, such as solar panels or a backup generator, to power essential systems like lighting, ventilation, and communication.

2. Container-Based Safe Spaces

For a more affordable alternative to underground bunkers, consider using shipping containers as safe spaces. Shipping containers can be buried or fortified above ground to create a secure shelter.

- Advantages: Lower cost and easier installation compared to a fully underground bunker. Shipping containers are also widely available and durable.

- Limitations: Shipping containers may require additional insulation and reinforcement to provide adequate protection.

Emergency Escape Routes: Planning for Quick and Safe Evacuations

In the event of a break-in, fire, or other emergency, having well-planned emergency escape routes can save lives. These routes allow you to exit your home quickly and discreetly if staying inside is no longer safe.

1. Creating Escape Plans for Every Room

Develop multiple escape plans for each room in your home. These should include primary and secondary exits, such as doors, windows, or hidden tunnels.

- Window Escapes: Equip bedrooms and living areas with escape ladders or wide enough windows to use as exits in case doors are blocked.

- Hidden Exits: In homes with secret passages or tunnels, ensure these exits are accessible and easy to navigate under stress.

Practical Tip: Practice escape routes with your family to ensure everyone knows what to do in an emergency. Time these drills to identify and eliminate any potential obstacles.

2. Discreet Exit Points

In high-risk situations, such as home invasions, it may be necessary to leave your home without being seen. Discreet exit points can include hidden doors, underground tunnels, or exits concealed within storage rooms or closets.

- Concealed Doors: Install hidden doors that blend with your home's design to provide an unnoticed escape route.

- Underground Tunnels: For maximum security, create underground tunnels that lead to a distant exit on your property or in a nearby wooded area.

Redundant Security Systems: Building Backup Defenses

Having redundant security systems in place ensures that your home remains secure even if the primary system fails. Whether due to power outages, system malfunctions, or sabotage, these backup measures provide an extra layer of defense.

1. Backup Alarms and Locks

In addition to your primary security system, install backup alarms and locks that don't rely on electricity or complex technology. Mechanical locks, battery-powered alarms, and manual security measures ensure your home stays secure during emergencies.

- Battery-Powered Alarms: Install battery-operated alarms on doors and windows that will function even if your main system goes down.

- Mechanical Locks: Use deadbolts and manual locks that don't rely on electronic systems, making them more resilient in a power outage or system failure.

2. Secondary Cameras and Motion Detectors

Set up a secondary set of security cameras and motion detectors that operate on a separate power source. This ensures that if one system fails, the other remains operational.

- Solar-Powered Security Cameras: Install cameras with built-in solar panels that can continue to function even during power outages.

- Motion-Activated Lights: Use motion-activated floodlights around your home's perimeter to detect and deter intruders even if your primary system goes down.

24. Security, Protecting Your Off-Grid Home from Potential Threats

Security is an essential aspect of off-grid living, as self-reliance extends to protecting your property, resources, and loved ones from potential intruders or wildlife. By establishing a comprehensive security system that includes visual deterrents, defensive systems, traps and weapons, and well-trained guard animals, you can create a secure environment that helps safeguard your home. This section covers strategies ranging from simple deterrents to advanced technology and legal considerations for using defensive tools.

Defensive Deterrents and Signs: Preventing Threats Before They Happen

One of the most effective ways to protect your off-grid home is through the use of defensive deterrents—strategies designed to dissuade intruders or unwanted visitors before they even attempt entry. These visual and physical deterrents serve as an early line of defense, making your home appear difficult to approach or breach.

1. The Power of Visual Deterrents

A well-placed sign or visible security feature can make intruders think twice about targeting your property. Most criminals seek easy, low-risk opportunities, so creating an environment that looks secure and well-monitored will often deter threats without the need for physical confrontation.

A. Warning Signs

Warning signs are a simple yet powerful tool in any security plan. Whether warning about dogs, cameras, or armed occupants, these signs send a clear message that the property is not an easy target.

Types of Signs:

- "Beware of Dog": Even if you don't have a dog, placing these signs can make intruders think twice.

- Surveillance Warnings: Signs that state the property is under 24-hour surveillance can discourage trespassing.

- Private Property: Marking your land as private and explicitly prohibiting trespassing is both a legal requirement in some places and a useful deterrent.

Practical Tip: Place warning signs at all entry points, including driveways, gates, and hidden access paths. Use highly visible, weather-resistant materials for long-lasting durability.

B. Visible Security Cameras

Even if your cameras are non-operational or simply decoys, visible security cameras can be an effective deterrent. Intruders are much less likely to target homes that appear to be monitored.

- Real vs. Decoy Cameras: Using a combination of both real and fake cameras can increase coverage and save on costs while still providing a deterrent effect. Place decoy cameras in prominent areas, and hide real cameras where they can capture important footage without being noticed.

Example: The Blink Outdoor Camera system is affordable, weatherproof, and motion-activated, making it ideal for remote off-grid properties.

2. Physical Deterrents

Physical barriers can deter threats by making entry physically challenging or intimidating. These barriers often require additional effort and tools to bypass, further discouraging unwanted visitors.

A. Motion-Activated Lights

Motion-activated lights are an excellent way to startle potential intruders and increase visibility around your property. These lights automatically switch on when movement is detected, drawing attention to any activity around your home and making it harder for someone to approach unnoticed.

- Best Locations: Install these lights near entry points such as doors, windows, and gates, as well as along driveways or walkways.

- Solar Options: Solar-powered motion lights are ideal for off-grid settings, as they don't require a wired power source.

Example: The Ring Solar Floodlight provides bright, motion-activated lighting powered entirely by solar energy.

B. Perimeter Alarms

Perimeter alarms create an audible alert when someone crosses a boundary around your property. These alarms can be placed at entry points, gates, or driveways to warn you of anyone approaching, giving you time to respond before they reach your home.

- Practical Tip: Use wireless, battery-powered perimeter alarms that don't rely on grid electricity, ensuring they function reliably in off-grid settings.

Setting Up Defensive Systems: Alarms, Cameras, and Motion Detectors

Incorporating defensive systems like alarms, cameras, and motion detectors enhances your security and provides critical real-time alerts to potential breaches. These technology-driven systems are invaluable for monitoring and protecting your off-grid property.

1. Security Alarms: Early Detection Systems

Security alarms alert you to potential intruders by sounding an alarm or sending notifications when a door, window, or other entry point is breached. In off-grid homes, where law enforcement or neighbors may be far away, alarms serve as your first line of defense, giving you critical time to react.

Types of Alarms

- Door and Window Alarms: These simple systems trigger when doors or windows are opened unexpectedly.

- Glass Break Sensors: These alarms detect the sound of breaking glass and can notify you of a forced entry attempt.

116

- Panic Alarms: Installed in key areas of your home, panic alarms allow you to trigger a loud alarm manually in case of emergency.

Practical Tip: Opt for solar-powered or battery-operated alarms that work even during power outages, ensuring continuous protection.

2. Surveillance Cameras: Monitoring Your Property

Surveillance cameras provide a real-time feed of what's happening around your home, allowing you to monitor activities even when you're away. In an off-grid setting, remote access and solar-powered systems are crucial for effective long-term monitoring.

- Remote Viewing: Many modern cameras offer remote access, allowing you to check live footage via a smartphone or computer.

- Solar-Powered Cameras: Look for solar-powered cameras that don't require an external power source, which makes them ideal for off-grid locations.

Example: The Reolink Argus 3 Pro is a solar-powered camera with night vision, two-way audio, and app-based remote monitoring.

3. Motion Detectors: Advanced Intrusion Detection

Motion detectors can be used indoors and outdoors to detect movement in key areas around your home. These detectors activate lights, alarms, or cameras when they sense motion, giving you an early warning of potential intrusions.

- Outdoor Motion Detectors: Place these detectors around gates, driveways, and windows to cover the most likely entry points.

- Indoor Motion Detectors: Install them in hallways or rooms that are not regularly used to catch intruders who have made it inside.

Traps and Weapons: Legal and Safe Defensive Tools

For off-grid security, some homeowners may consider using traps and weapons as defensive tools. However, it's essential to understand the legal and ethical considerations before setting up any potentially harmful devices.

1. Non-Lethal Deterrents

Non-lethal deterrents are an effective way to discourage intruders without causing permanent harm. These can include traps that disable or frighten but don't cause fatal injuries, making them more legally acceptable in most areas.

A. Tripwires and Noisemakers

Tripwires connected to noisemakers or alarm systems are simple yet effective ways to alert you when someone is moving through your property. When tripped, these devices create a loud noise that startles the intruder and warns you of their presence.

- Best Use: Place tripwires near access points like driveways, gates, or around the perimeter of sensitive areas.

- Practical Tip: Use reflective tape or paint to make the tripwires less visible at night.

B. Pepper Spray Traps

Pepper spray traps can be set up in entryways to release a burst of pepper spray when triggered. These traps provide a strong deterrent, incapacitating intruders without causing permanent harm.

Legal Tip: Always check local regulations regarding the use of non-lethal traps, as laws may vary by jurisdiction.

2. Firearms for Home Defense

For those who are comfortable and legally permitted, firearms can be an important tool for home defense. However, owning and using firearms requires strict attention to safety, storage, and legal responsibilities.

A. Safe Storage of Firearms and Ammunition

Safe storage is paramount to prevent accidents or theft. Firearms should always be stored securely, and access should be limited to authorized individuals only.

- Gun Safes: Invest in a high-quality, fireproof gun safe that can securely store your firearms and ammunition. Safes with biometric or keypad locks provide quick access in emergencies.

- Ammunition Storage: Store ammunition separately from firearms in a secure, dry location to prevent deterioration and unauthorized access.

B. Legal Considerations for Firearms

Before acquiring firearms, ensure that you understand the laws in your area regarding firearm ownership, usage, and self-defense. Laws vary significantly, so knowing what's legal and permissible is critical to avoid legal consequences.

Practical Tip: Always complete a certified firearms safety course to ensure you know how to handle and store your firearm responsibly.

Training for Security: Self-Defense and Firearm Safety

Training is a crucial component of any security plan, particularly when firearms or other defensive tools are involved. Knowing how to properly use your defensive tools and practicing self-defense techniques ensures that you're prepared to act effectively in an emergency.

1. Basic Self-Defense Training

Even without weapons, basic self-defense training can be invaluable. Learning how to physically defend yourself and disarm an attacker gives you confidence and control in dangerous situations.

- Simple Techniques: Focus on easy-to-learn, effective moves like blocking, striking, and escaping holds. Many self-defense courses emphasize situational awareness and de-escalation strategies.

- Training Options: Consider enrolling in local self-defense classes or practicing with a partner to build skills and confidence.

2. Firearms Training and Safety

If you own firearms for home defense, it's essential to receive proper training on how to use, maintain, and store them safely.

Firearms Safety: Learn the four basic rules of firearm safety:

1. Treat every firearm as if it's loaded.

2. Never point a firearm at anything you don't intend to shoot.

3. Keep your finger off the trigger until you're prepared to fire.

4. Know your target and what's beyond it.

- Practice: Regular practice at a shooting range is essential for maintaining accuracy and confidence in using your firearm.

Practical Tip: Invest in gun safety classes and practice regularly with any weapons you keep for defense to ensure you are comfortable and capable of using them under stress.

Guard Animals: Nature's Home Security

Guard animals are a natural, effective way to enhance the security of your off-grid home. Well-trained dogs, geese, or other animals can provide an early warning system and even physical defense against intruders.

1. Guard Dogs: Loyal Protectors

Guard dogs are one of the most trusted security animals, capable of alerting you to intruders and, if necessary, defending your property. The key to an effective guard dog is proper training and socialization.

- Choosing a Guard Dog Breed: Certain breeds, such as German Shepherds, Rottweilers, and Mastiffs, are known for their guarding instincts and loyalty. Choose a breed that suits your home's environment and your ability to train and care for it.

- Training Your Dog: Guard dogs must be trained to differentiate between regular visitors and potential threats. Basic obedience training is a must, followed by more advanced protection training.

Practical Tip: Dogs require daily care and attention, so only choose this option if you're ready for the responsibility of having a pet with dual roles as a companion and protector.

2. Geese: Natural Alarms

Surprisingly, geese are highly effective guard animals. Known for their territorial nature and loud honking, geese can quickly alert you to any unusual activity on your property.

- Advantages: Geese are low-maintenance and don't require specialized training to act as a natural alarm system. They're also known for their keen senses and ability to detect movement from a distance.

- Practical Use: Keep geese in fenced areas where they can patrol and alert you to intruders.

Conclusion: Building a Comprehensive Security Plan for Off-Grid Living

Creating a secure off-grid home involves a combination of visual deterrents, defensive systems, traps, and well-trained animals. By using technology like alarms and cameras, implementing self-defense strategies, and training guard animals, you can protect your home and family from external threats. Whether you prefer non-lethal deterrents or rely on firearms for defense, the key to effective security is preparation, training, and responsible use of tools and resources.

Module G | Food Sustainability - From Production to Preservation

25. Producing Food: Sustainable Methods for Off-Grid Living

Producing your own food is at the heart of self-sufficiency, especially when living off-grid. Whether you have access to fertile land, limited space, or even water resources, a variety of sustainable techniques can help you grow and raise food to meet your nutritional needs. In this section, we will explore different approaches such as gardening with permaculture, aquaponics and hydroponics, and animal husbandry, each offering unique benefits for off-grid living.

Gardening and Permaculture: Growing Abundant Food Sustainably

Permaculture is a design philosophy that mimics the patterns and relationships found in natural ecosystems to create sustainable and self-sufficient food production systems. It goes beyond traditional gardening by considering long-term resilience, biodiversity, and minimal inputs.

1. Understanding Permaculture Principles

At its core, permaculture seeks to work with nature rather than against it. This approach to gardening emphasizes regenerative practices, meaning your garden can improve soil quality, conserve water, and increase biodiversity without depleting resources. Here are the key principles of permaculture gardening:

A. Observe and Interact

Start by observing the natural characteristics of your land. Pay attention to sunlight patterns, wind direction, water flow, and existing vegetation. This will help you design a system that takes advantage of the natural conditions on your property.

- Practical Tip: Spend time in different parts of your property at various times of day to fully understand how sunlight, shade, and wind affect each area.

B. Design for Diversity and Stability

Diverse ecosystems are more resilient. In permaculture, polycultures (growing different crops together) replace traditional monocultures, ensuring that each plant supports others in some way. This not only reduces the risk of pests and diseases but also creates a more stable food supply.

- Example: The classic "Three Sisters" companion planting of corn, beans, and squash is a well-known permaculture technique. The corn provides support for the beans, which fix nitrogen in the soil, while the squash shades the ground, reducing evaporation.

C. Use Natural Resources Wisely

Permaculture encourages careful management of resources such as water and soil. For example, rainwater can be captured in tanks or directed to garden beds using swales—ditches designed to capture and distribute water efficiently.

- Practical Tip: Mulch your garden beds to retain moisture and build up soil organic matter, which improves water retention and reduces the need for frequent irrigation.

121

2. Setting Up a Permaculture Garden

Starting a permaculture garden requires thoughtful planning, but once established, it becomes more self-sustaining over time. Here's a step-by-step guide to get you started:

A. Zoning and Layout

Permaculture uses a zone system to organize your space based on how frequently each area needs to be accessed. The closer a zone is to your home, the more interaction it requires.

- Zone 1: This is the area closest to your home, where you grow herbs, vegetables, and small plants that require daily attention.
- Zone 2: Further out, this area is for perennial crops, larger vegetable patches, and plants that need less frequent care.
- Zone 3: Typically reserved for orchard trees, larger-scale crops, and grazing areas for livestock.

B. Building Healthy Soil

Healthy soil is the foundation of any successful garden. In permaculture, the goal is to build soil health through organic matter and biodiversity, reducing the need for chemical fertilizers. Composting, vermiculture (worm farming), and mulching are essential techniques for improving soil quality.

- Practical Tip: Create raised beds or hugelkultur mounds, which are layered garden beds filled with decaying wood and organic matter. These beds hold moisture and improve soil fertility over time.

C. Plant Selection and Polycultures

Choose plants that are well-suited to your climate and that complement each other when grown together. Native plants are often more resilient and require less maintenance.

- Example: In dry climates, grow drought-tolerant plants like amaranth, quinoa, and chard. Pair them with deep-rooted perennials that help break up compacted soil and retain water, such as comfrey or yarrow.

3. Long-Term Sustainability in Permaculture

One of the key advantages of permaculture is its emphasis on perennial plants—those that live for several years and require less maintenance once established. By integrating these into your garden, you reduce the need for annual planting and create a more stable food system.

A. Perennial Vegetables and Fruits

Perennials like asparagus, rhubarb, and artichokes can provide a continuous harvest year after year with minimal effort. Fruit trees like apples, pears, and figs can also be incorporated into the system, along with berry bushes like blueberries or raspberries.

B. Forest Gardens

A forest garden is a permaculture design that mimics the structure of a natural forest, with multiple layers of vegetation, from tall canopy trees to low-growing ground covers. In this system, every layer provides food or resources, and the diversity helps protect against pests and diseases.

- Practical Tip: Start with a few key fruit and nut trees as the backbone of your forest garden, then add shrubs, herbs, and ground covers in layers beneath them.

Example: A forest garden could include a canopy layer of chestnut or walnut trees, an understory of hazelnut bushes, and a ground cover of strawberries or mint.

Aquaponics and Hydroponics: Growing Food in Water-Based Systems

For those with limited space or less-than-ideal soil conditions, aquaponics and hydroponics provide innovative ways to grow food using water-based systems. These methods are particularly well-suited for off-grid living because they maximize efficiency, require minimal land, and produce a continuous supply of fresh vegetables and fish.

1. Hydroponics: Soil-Free Gardening

Hydroponics is a method of growing plants using nutrient-rich water instead of soil. Plants are supported in a growing medium like coconut coir, perlite, or vermiculite, while the roots are directly exposed to a nutrient solution that delivers essential minerals and elements.

A. Benefits of Hydroponics

- Space Efficiency: Hydroponic systems can be set up vertically or horizontally, making them ideal for small spaces like balconies, greenhouses, or rooftops.

- Water Conservation: Hydroponics uses up to 90% less water than traditional soil gardening because water is recirculated through the system.

- Faster Growth: Plants grown hydroponically often grow faster than those in soil because they have direct access to nutrients and oxygen.

B. Basic Hydroponic Systems

There are several types of hydroponic systems, ranging from simple DIY setups to more advanced commercial designs.

- Nutrient Film Technique (NFT): In this system, a thin film of nutrient-rich water flows over the roots of the plants, allowing them to absorb nutrients while still getting oxygen from the air.

- Deep Water Culture (DWC): Plants are suspended above a reservoir of nutrient solution, with their roots submerged in the water. This system is simple to build and maintain, making it great for beginners.

- Wick System: In this passive hydroponic system, a wick draws water and nutrients from a reservoir to the plant roots. It's low-maintenance but best suited for small plants like herbs or lettuce.

Practical Tip: Use LED grow lights or take advantage of natural sunlight to support your hydroponic system, especially if you're growing indoors.

2. Aquaponics: Combining Fish and Plants for a Symbiotic System

Aquaponics is an innovative system that integrates aquaculture (fish farming) with hydroponics (soilless plant cultivation). In this setup, fish waste supplies essential nutrients for the plants, while the plants, in turn, help purify and filter the water, establishing a mutually beneficial relationship between the two components.

A. How Aquaponics Works

In an aquaponics system, fish are kept in a tank where they produce waste. This waste is rich in nutrients, particularly nitrogen, which plants need to grow. The water from the fish tank is pumped into a grow bed where plants absorb these nutrients. The cleaned water is then recirculated back into the fish tank.

- Fish Species: Common fish for aquaponics include tilapia, trout, catfish, and koi, as they are hardy and can tolerate the fluctuating conditions in small-scale systems.
- Plants: Aquaponics works well for leafy greens like lettuce, kale, spinach, and herbs, as well as fruiting plants like tomatoes and peppers.

B. Benefits of Aquaponics

- Sustainable Food Production: Aquaponics provides both plant and fish harvests, making it a highly efficient way to produce food.
- Water Efficiency: Like hydroponics, aquaponics recirculates water, making it more efficient than traditional soil gardening.
- Minimal Inputs: The only regular input required is fish feed, which can be supplemented with homegrown food like worms or insect larvae.

C. Setting Up an Aquaponic System

To set up an aquaponic system, you'll need a fish tank, a water pump, grow beds, and a filtration system.

- Fish Tank: Choose a tank that's large enough to support the number of fish you plan to raise. A 200-gallon tank is sufficient for a small backyard system.
- Grow Beds: Place grow beds above the fish tank, filled with a growing medium like clay pebbles or gravel.
- Water Circulation: Use a water pump to circulate the nutrient-rich water from the fish tank to the grow beds, where plants absorb the nutrients before the water is returned to the tank.

Practical Tip: Keep the pH levels in your aquaponic system balanced (between 6.8 and 7.2) to ensure that both the fish and plants thrive.

Animal Husbandry: Raising Livestock for Sustainable Food Production

Raising animals is another key aspect of off-grid food production. Animal husbandry allows you to produce eggs, milk, meat, and even natural fertilizers from animals like chickens, goats, and other livestock.

1. Chickens: The Perfect Small-Scale Livestock

Chickens are one of the most popular animals to raise for off-grid food production due to their ability to provide both eggs and meat. They're relatively low-maintenance, require little space, and contribute to pest control by eating insects.

A. Setting Up a Chicken Coop

A well-designed chicken coop provides shelter, protection from predators, and a safe place for chickens to lay eggs.

- Size: Each chicken needs about 2-3 square feet inside the coop and at least 8-10 square feet in an outdoor run.

- Nesting Boxes: Provide one nesting box for every 3-4 chickens, filled with straw or wood shavings for comfort.

- Predator Protection: Secure the coop with strong wire mesh to keep out predators like raccoons, foxes, and hawks.

B. Feeding and Care

Chickens can be fed a combination of commercial feed, kitchen scraps, and free-range foraging. Free-ranging chickens can help control pests and improve the soil through their natural scratching and manure.

- Supplements: Provide chickens with grit and calcium supplements (like crushed oyster shells) to aid digestion and strengthen eggshells.

2. Goats: Versatile Livestock for Milk, Meat, and Land Management

Goats are an excellent choice for off-grid homesteads, offering milk, meat, and natural land clearing. They're hardy animals that thrive in a variety of climates and can forage on a wide range of plants.

A. Choosing a Goat Breed

Different breeds of goats are suited to different purposes, so choose one based on your needs.

- Dairy Breeds: Nubian and Saanen goats are known for their high milk production.

- Meat Breeds: Boer and Kiko goats are popular meat breeds, known for their fast growth and tender meat.

B. Housing and Fencing

Goats need secure shelter and fencing to protect them from predators and to keep them contained.

- Shelter: A simple three-sided shelter is usually enough to protect goats from the elements. Ensure it's well-ventilated and dry.

- Fencing: Goats are notorious escape artists, so invest in strong, high fencing (at least 4-5 feet tall) to keep them contained.

3. Other Livestock Options for Off-Grid Living

In addition to chickens and goats, other animals can be valuable on an off-grid homestead, depending on your space and resources.

Rabbits

Rabbits are easy to raise and provide a sustainable source of meat. They require minimal space and can be fed on a diet of grasses, hay, and kitchen scraps.

- Practical Tip: Rabbit manure is an excellent fertilizer that can be added directly to garden beds without composting.

Ducks

Ducks are similar to chickens in terms of care, but they're better suited to wet environments. They provide eggs, meat, and can help control garden pests like slugs and snails.

Conclusion: Producing Food Sustainably Off the Grid

Whether you choose to grow your food through permaculture gardening, adopt aquaponics or hydroponics, or raise livestock like chickens and goats, off-grid food production offers a sustainable way to provide for your family. By using these techniques, you can create a resilient and self-sufficient food system that works in harmony with nature, minimizing inputs while maximizing yields. Each method can be adapted to suit your land, climate, and personal preferences, ensuring that your off-grid home remains productive and food-secure year-round.

26. Procuring Food: Mastering Nature's Resources for Off-Grid Living

When living off the grid, procuring food from your natural environment becomes an essential skill. While growing your own food is a sustainable option, knowing how to forage, hunt, and fish provides the flexibility and resilience needed to thrive without relying on external sources. This chapter focuses on these three key skills, offering practical guidance for identifying wild edibles, hunting for sustenance, and fishing to supplement your diet.

Foraging: Harvesting Nature's Bounty

Foraging is the practice of gathering wild plants, fruits, nuts, and fungi for food. This age-old skill is both practical and rewarding, allowing you to tap into nature's abundant resources. However, it requires knowledge and care to ensure that what you gather is safe and sustainable.

1. Identifying Edible Plants: What to Look For

Learning to identify edible plants is the first step in foraging. While some wild plants are safe and nutritious, others can be toxic or harmful. It's crucial to develop a solid understanding of local flora to avoid dangerous mistakes.

A. Common Edible Plants

There are many edible plants found in various environments across the world. Here are a few that are widely available and easy to identify:

- Dandelion (Taraxacum officinale): The entire plant is edible, including the leaves, flowers, and roots. Dandelion leaves are rich in vitamins A and C, while the roots can be roasted to make a caffeine-free coffee substitute.

- Chickweed (Stellaria media): This small, delicate plant is often found in disturbed soil or along garden edges. The leaves and stems can be eaten raw in salads or cooked in soups and stews.

- Wild Garlic (Allium ursinum): Easily identifiable by its strong garlic smell, wild garlic grows in damp woodlands. The leaves, flowers, and bulbs are all edible and can be used to flavor a variety of dishes.

- Stinging Nettle (Urtica dioica): Nettles must be cooked to neutralize their sting, but they are rich in iron, vitamins, and minerals. They can be used in soups, teas, or as a spinach substitute.

Practical Tip: Carry a foraging guidebook specific to your region or use a plant identification app to help you safely identify plants in the wild.

B. Poisonous Plants to Avoid

While foraging can be rewarding, it also carries risks. Some wild plants closely resemble edible species but are highly toxic. Familiarize yourself with the following dangerous plants to avoid.

- Poison Hemlock (Conium maculatum): This plant resembles wild carrot or parsley but is extremely poisonous. It has purple-spotted stems and finely divided leaves. Even a small amount can be fatal if ingested.

- Deadly Nightshade (Atropa belladonna): Found in woodlands and scrub, deadly nightshade has dark purple berries that resemble edible species but contain toxic alkaloids. The plant is highly poisonous.

- Foxglove (Digitalis purpurea): Known for its tall, spiky flowers, foxglove is toxic to humans and can cause heart issues if consumed.

Practical Tip: Follow the Universal Edibility Test if you're unsure about a plant. Start by rubbing the plant on your skin and lips, waiting for any reactions. If none occur, taste a small piece and wait several hours before consuming more.

2. Safe Foraging Practices: Harvesting Without Harming

Foraging is about more than just gathering food—it's also about respecting nature and ensuring that wild plant populations remain healthy and abundant for future use.

A. Harvesting Sustainably

Sustainable foraging involves taking only what you need and ensuring that you don't deplete the plant population in a given area. Over-harvesting can lead to the decline of certain species, disrupting local ecosystems.

- Practical Tip: Follow the "Rule of Thirds": Harvest no more than one-third of any plant or patch, leaving the rest to regenerate. For rare or endangered species, avoid harvesting altogether.

B. Respecting the Environment

While foraging, be mindful of your surroundings and minimize your impact on the environment. Avoid trampling on plants or disturbing wildlife habitats, and always leave the area as you found it.

- Practical Tip: Use biodegradable bags or baskets to carry your foraged items. Plastic bags can damage delicate plants and contribute to pollution.

Hunting: Gathering Protein from the Wild

Foraging for plants is essential, but hunting provides a critical source of protein, which is often harder to obtain in the wild. Hunting for sustenance requires skill, patience, and adherence to ethical practices to ensure both safety and sustainability.

1. Understanding the Basics of Hunting

Hunting involves tracking and harvesting wild animals for food. To do so effectively, you need to understand the habits and habitats of your prey, how to track them, and how to process the meat safely.

A. Common Game Animals

Depending on your location, various game animals can be hunted for food. Some of the most common species include:

- Deer: In many areas, deer are abundant and provide a significant amount of meat. Their habits vary by region, but they are often found near water sources and forested areas.

- Rabbits: Rabbits are small, quick, and reproduce rapidly, making them a reliable source of protein. They are commonly found in meadows, brushy areas, and forest edges.

- Wild Boar: These animals are larger and more dangerous but provide a substantial amount of meat. They can be found in forests, grasslands, and agricultural areas.

- Squirrels: Squirrels are often overlooked, but they can be a good source of food, especially in forested areas. They are usually active during the day and can be found near trees.

B. Tools for Hunting

When hunting, choosing the right tools is crucial for both effectiveness and safety. Depending on the game and local laws, different types of weapons may be suitable:

- Firearms: Rifles, shotguns, and handguns are common tools for hunting larger game like deer or boar. Ensure that you are familiar with firearm safety and local hunting regulations before using these weapons.

- Bows and Crossbows: Bows offer a quieter, more stealthy option for hunting, ideal for areas where you want to avoid drawing attention or disturbing the environment.

- Traps and Snares: For smaller game like rabbits or squirrels, traps and snares can be an effective and low-energy way to hunt. However, it's important to check these traps regularly to ensure humane treatment of the animals.

2. Hunting Techniques: Tracking, Stalking, and Trapping

Hunting requires not only the right tools but also the right techniques. Knowing how to track, stalk, and trap game will improve your chances of a successful hunt.

A. Tracking Game

Tracking is a critical skill for any hunter. Learn to identify animal tracks, scat, and feeding signs to determine where animals are likely to be found.

- Footprints: Each animal has a distinctive footprint. Look for deer tracks, which have cloven hooves, or rabbit tracks, which show larger hind feet and smaller front feet.

- Droppings: Animal droppings, or scat, can tell you what animals are nearby and how recently they passed through the area. Fresh droppings indicate that the animal is close, while older droppings suggest they've moved on.

Practical Tip: Soft ground near water sources is ideal for finding tracks. Pay attention to trails that lead to water, as animals often follow these paths.

B. Stalking and Ambush

Once you've identified the presence of game, the next step is to stalk or ambush them. This involves approaching quietly and positioning yourself in a spot where the animal is likely to come within range.

- Wind Direction: Always keep the wind in your face when stalking game. Animals have an excellent sense of smell and can detect you long before you see them if the wind is blowing your scent toward them.

- Stealth: Move slowly and quietly, avoiding sudden movements that can startle your prey. Use natural cover like trees, rocks, and bushes to conceal yourself as you approach.

C. Setting Traps and Snares

Traps and snares are ideal for small game hunting, particularly when you want to conserve energy. However, they require skill and practice to set up correctly.

- Snares: Snares are loops made of wire or strong cord that tighten around an animal's neck or body when triggered. Place snares along animal trails or in areas where you've seen tracks.

- Deadfall Traps: These traps use a heavy object, like a rock or log, to crush the animal when triggered. Deadfall traps are effective for small mammals like rabbits or squirrels.

3. Processing Game: From Field to Table

Once you've successfully hunted game, knowing how to process the meat is crucial to ensure it's safe to eat and can be preserved for future use.

A. Field Dressing

Field dressing is the process of removing the internal organs from the animal as soon as possible after the kill. This prevents the meat from spoiling and makes it easier to transport.

Steps to Field Dress a Deer:

1. Lay the animal on its back and make a small incision just below the ribcage.
2. Carefully cut through the skin and muscle without puncturing the internal organs.
3. Remove the organs, starting with the stomach and intestines, followed by the heart and lungs.
4. Rinse the cavity with clean water and allow it to drain before transporting the animal.

B. Butchering

Once you've brought the animal back to your homestead, you'll need to butcher it into usable cuts of meat. Larger animals like deer and boar should be broken down into steaks, roasts, and ground meat, while smaller animals like rabbits can be cooked whole or in parts.

- Practical Tip: Invest in quality knives and butchering tools to make the process easier. A bone saw, fillet knife, and meat grinder are essential tools for processing game at home.

Fishing: Catching and Harvesting Fish for Food

Fishing is another crucial skill for off-grid living, providing a consistent source of protein. Whether you're fishing in rivers, lakes, or oceans, knowing how to catch, clean, and preserve fish can significantly enhance your food supply.

1. Fishing Techniques: Hooks, Nets, and Traps

Fishing can be done in a variety of ways, depending on the body of water and the type of fish you're targeting.

A. Rod and Reel Fishing

This is the most common method of fishing and involves using a fishing rod and reel to cast a line with bait or lures into the water.

- Bait: Live bait like worms, minnows, or crickets work well for attracting fish. Artificial lures can also be effective, depending on the species you're targeting.

- Casting: Cast your line into areas where fish are likely to be feeding, such as near rocks, fallen trees, or underwater vegetation.

Practical Tip: Use a bobber to keep your bait at the right depth and to alert you when a fish bites.

B. Setting Fish Traps and Nets

Fish traps and nets allow you to catch fish passively, which is particularly useful if you want to catch large quantities or free up time for other tasks.

- Fish Traps: These are cages or enclosures that allow fish to swim in but not out. Place traps in shallow water near the shore, where fish are likely to pass through.

- Nets: Gill nets and seine nets are used to catch fish in larger bodies of water. They are especially effective for catching multiple fish at once.

C. Spearfishing

For those living near shallow, clear waters, spearfishing can be an effective and low-cost way to catch fish. It requires patience, accuracy, and the ability to stay underwater for short periods.

2. Cleaning and Processing Fish

After catching fish, you need to clean and process them quickly to ensure the meat remains fresh.

A. Scaling and Gutting

The first step in processing a fish is to remove the scales and internal organs.

- Scaling: Use a fish scaler or the back of a knife to remove the scales by scraping from the tail toward the head.

- Gutting: Make an incision along the belly of the fish, from the anus to the gills. Remove the internal organs and rinse the cavity with clean water.

B. Filleting

For larger fish, filleting allows you to separate the meat from the bones, making it easier to cook and store.

Steps to Fillet:

1. Lay the fish on its side and make a cut just behind the gills, down to the backbone.

2. Slide the knife along the backbone, cutting through the ribs, to separate the fillet from the body.

3. Repeat on the other side, then remove any remaining bones.

3. Preserving Fish: Smoking, Drying, and Canning

Once you've caught and processed your fish, you'll need to preserve it for long-term storage, especially if you're fishing in large quantities.

Smoking Fish

Smoking is a traditional method of preserving fish that not only extends its shelf life but also adds a rich, smoky flavor.

- Hot Smoking: Fish is cooked slowly at a low temperature over smoldering wood. This method both cooks and preserves the fish.

- Cold Smoking: The fish is exposed to smoke at a lower temperature without cooking it, preserving the fish for long-term storage.

Drying Fish

Drying fish involves removing the moisture to prevent spoilage. You can do this by air-drying or using a dehydrator.

Practical Tip: Salt the fish before drying to help draw out moisture and improve preservation.

Canning Fish

For longer-term storage, canning is an excellent option. Pack the fish into sterilized jars with salt, then process them in a pressure canner to ensure they are sealed and preserved safely.

Conclusion: Procuring Food in the Wild

Foraging, hunting, and fishing are invaluable skills for off-grid living, allowing you to procure food directly from nature. By learning to identify edible plants, mastering hunting techniques, and becoming proficient in fishing, you can create a sustainable and diverse food supply that reduces your reliance on external resources. Whether you're harvesting wild greens, setting traps for game, or catching fish, these skills will help you thrive in a self-sufficient lifestyle.

27. Preparing Food: Off-Grid Cooking Techniques for Sustainable Living

Cooking off the grid requires a blend of traditional methods and modern innovations that allow you to prepare meals without relying on electricity or gas. By harnessing the power of fire, using solar ovens, mastering the use of rocket stoves, and relying on wood stoves, you can create delicious meals while minimizing energy consumption. This section explores these off-grid cooking techniques in detail, providing practical advice on how to use each method effectively.

Cooking with Fire: The Oldest Method of Off-Grid Cooking

Cooking over an open fire is one of the most primitive and reliable methods of preparing food. Whether using a campfire or a fire pit, this technique allows you to cook almost anything, from boiling water to roasting meats.

1. Setting Up a Safe and Efficient Cooking Fire

Creating a fire for cooking requires a few essential steps to ensure safety and efficiency. A well-constructed fire burns steadily and provides consistent heat for cooking.

A. Building a Fire Pit

A fire pit is an essential feature for safe, controlled cooking. If you don't already have one, here's how to create a basic fire pit for off-grid cooking:

- Dig a Shallow Pit: Choose a flat, clear area, away from trees, structures, or anything flammable. Dig a shallow pit about 1-2 feet deep and line it with stones or bricks.

- Clear the Area: Remove any debris or vegetation around the fire pit to prevent accidental fires. A radius of at least 10 feet around the fire should be clear.

- Wind Protection: Use rocks or a simple windbreak to protect the fire from gusts, ensuring it burns steadily.

B. Starting and Maintaining a Cooking Fire

Once your fire pit is ready, the next step is to build and maintain the fire for cooking:

- Kindling and Fuel: Start with small, dry kindling like twigs, leaves, or paper. Once it's burning steadily, gradually add larger pieces of wood or logs.

- Fire Layout for Cooking: Use the teepee method (wood arranged in a conical shape) for a quick, hot fire or the log cabin method (stacking wood in a square shape) for a longer, steadier burn, perfect for slower cooking methods like simmering or roasting.

- Managing Heat: Control the heat by adjusting the size of the fire or moving the food closer or farther from the flames. If you need more heat, add more wood. For less heat, spread out the coals and reduce the flame.

Practical Tip: A grill grate or a simple rack can be placed over the fire to hold pots, pans, or even food directly.

2. Off-Grid Cooking Techniques Using Fire

There are many ways to cook over a fire, each offering different levels of control and flavor. Here are some of the most common methods:

A. Boiling

Boiling is one of the simplest and most effective methods of cooking over a fire. All you need is a pot, water, and a steady flame.

- Best Uses: Boiling is ideal for cooking grains like rice, pasta, or potatoes, as well as for making soups, stews, or boiling water for purification.

- Practical Tip: Use a tripod or hang the pot from a stick across the fire for greater control over the heat.

B. Roasting and Grilling

Roasting meat or vegetables over a fire imparts a smoky flavor and cooks food quickly at high heat.

- Roasting on Skewers: Skewers or sticks can be used to hold food over the fire. This method works well for small cuts of meat, fish, or vegetables.

- Grilling on a Grate: Place a grill grate over the fire to cook larger pieces of meat or vegetables. Keep the food moving to prevent burning and ensure even cooking.

C. Baking in Hot Coals

For baking bread, potatoes, or other items, hot coals can be used as a natural oven.

- How to Bake: After your fire has burned down to hot coals, place the food wrapped in foil or placed inside a Dutch oven directly in the coals. Cover the top with more coals to evenly distribute heat.

- Example: Campfire Bread can be made by wrapping dough around a stick and slowly cooking it over the coals.

133

Practical Tip: Rotate food frequently when roasting or grilling to avoid burning on one side.

Solar Ovens: Harnessing the Sun's Energy for Cooking

Solar ovens are an eco-friendly and efficient way to cook off-grid by harnessing the sun's power. They use reflective surfaces to concentrate sunlight, generating enough heat to cook a wide variety of meals.

1. How Solar Ovens Work

Solar ovens rely on the greenhouse effect to cook food. Sunlight is reflected and focused into a dark, insulated box, where it is converted into heat. The trapped heat increases the temperature inside the oven, allowing you to cook food without any fuel.

Types of Solar Ovens

There are three main types of solar ovens, each with unique benefits:

- Box Ovens: These are insulated boxes with a glass or plastic cover to trap heat. They are versatile and can reach temperatures of up to 300°F (149°C), suitable for baking, roasting, and slow cooking.

- Parabolic Ovens: These ovens use a curved, reflective dish to focus sunlight on a single point. They reach higher temperatures (up to 450°F or 232°C) and are ideal for faster cooking tasks like frying or grilling.

- Panel Cookers: These use multiple flat reflective panels to direct sunlight toward the cooking pot. They are simple to build but may not reach as high temperatures as box or parabolic ovens.

Practical Tip: A thermometer inside the oven helps monitor the cooking temperature and ensures the food is cooked safely.

2. Cooking Techniques with Solar Ovens

Cooking with a solar oven is similar to using a slow cooker or a traditional oven, but it requires a bit more patience and attention to weather conditions.

A. Baking in a Solar Oven

Solar ovens are perfect for baking bread, cakes, and other baked goods. The even, steady heat cooks food slowly, producing tender and flavorful results.

- Baking Bread: Place dough in a dark baking pan (dark colors absorb more heat) and set it inside the solar oven. Baking times will be longer than with a conventional oven, typically taking 2-3 hours depending on the sunlight and temperature.

- Practical Tip: Rotate the solar oven throughout the day to follow the sun, ensuring even heating.

B. Roasting Meat and Vegetables

The slow, gentle heat of a solar oven makes it excellent for roasting meat and vegetables, retaining moisture while cooking evenly.

- Roasting Meat: Season meat as you would for conventional roasting and place it in a covered pot inside the oven. Chicken, lamb, and pork all roast beautifully in a solar oven, typically taking 3-6 hours.

- Practical Tip: For extra flavor, marinate the meat overnight before placing it in the solar oven.

C. Cooking Grains and Legumes

Solar ovens are particularly useful for slow-cooking grains like rice or quinoa and legumes like beans or lentils.

- Best Practices: Use a pot with a lid to retain moisture, and add extra water since cooking times are longer. Grains and legumes may take 2-4 hours to cook fully.

Practical Tip: Monitor the weather. Cloudy or overcast conditions will slow cooking times significantly.

Rocket Stoves: Efficient Cooking with Minimal Fuel

Rocket stoves are an energy-efficient way to cook using small amounts of wood or biomass. Designed to maximize heat output while minimizing fuel consumption, these stoves are perfect for off-grid cooking, especially in areas where wood is abundant but you want to use as little as possible.

1. How Rocket Stoves Work

A rocket stove uses an insulated vertical combustion chamber that creates a strong draft, drawing air through the stove and causing the wood to burn hotter and more efficiently. The high heat produced by this process allows you to cook quickly using just a handful of small sticks or twigs.

Advantages of Rocket Stoves

- Fuel Efficiency: Rocket stoves burn fuel much more efficiently than open fires, requiring significantly less wood or biomass to cook the same amount of food.

- Heat Concentration: The design of the stove focuses heat on the cooking surface, reducing cooking times.

- Portability: Many rocket stoves are small and lightweight, making them easy to transport or use in multiple locations around your property.

2. Cooking with a Rocket Stove

Rocket stoves are versatile and can be used for a variety of cooking methods, including boiling, frying, and simmering.

Boiling Water or Cooking Grains

Rocket stoves are ideal for boiling water or cooking grains like rice, quinoa, or pasta. The intense heat boils water quickly, allowing you to cook grains in a fraction of the time it takes over a traditional fire.

Practical Tip: Keep a kettle or pot on the stove at all times when using the rocket stove, as it's an efficient way to ensure you always have hot water ready for cooking or cleaning.

Frying and Sautéing

Rocket stoves can reach high enough temperatures for frying or sautéing food. The concentrated heat allows you to fry meats, vegetables, or eggs quickly.

- Best Practices: Use a heavy-bottomed skillet to distribute the heat evenly and avoid burning food.

- Practical Tip: Keep a close eye on the food when frying or sautéing, as rocket stoves can heat very quickly.

Wood Stoves: A Classic Off-Grid Cooking Solution

For those living off the grid in colder climates, wood stoves offer both heating and cooking capabilities. Unlike campfires or open flames, wood stoves provide a controlled environment for slow cooking, baking, and simmering, making them a versatile option for preparing meals.

1. Types of Wood Stoves for Cooking

There are two main types of wood stoves used for cooking: traditional wood stoves and cookstoves with integrated ovens and cooking surfaces.

A. Traditional Wood Stoves

Traditional wood stoves are primarily used for heating, but many models feature flat tops that double as cooking surfaces. These stoves work well for boiling, frying, and simmering food.

- Best Uses: Use cast-iron pots and pans directly on the stove's surface to cook stews, soups, and other one-pot meals.

- Practical Tip: Keep a pot of water simmering on the stove at all times to add humidity to the air and ensure hot water is always available.

B. Cookstoves with Ovens

Cookstoves are wood-burning stoves specifically designed for cooking, featuring ovens, warming shelves, and multiple cooking surfaces. These stoves are ideal for off-grid homes, allowing you to bake, roast, and boil simultaneously.

- Best Uses: The oven in a cookstove works similarly to a conventional oven, allowing you to bake bread, roast meats, or even make pizza.

- Practical Tip: Use the stovetop to cook while the oven is in use, maximizing fuel efficiency by cooking multiple dishes at once.

2. Cooking Techniques with a Wood Stove

Wood stoves provide a steady, even heat for cooking, but they require careful management of the fire to maintain the right temperature.

A. Simmering and Slow Cooking

The gentle, steady heat of a wood stove makes it perfect for simmering stews, soups, and slow-cooking tougher cuts of meat.

- Best Practices: Use cast-iron or heavy-bottomed pots to retain heat and avoid scorching. Place the pot directly on the stove's surface or use a trivet for gentler heat.

- Practical Tip: Wood stoves can be unpredictable in temperature. Adjust the position of the pot or pan on the stove to regulate the heat as needed.

B. Baking in a Wood Stove Oven

If your wood stove has an oven, baking is a great way to make use of the steady, dry heat. From bread to casseroles, a wood stove oven can handle most baking tasks.

- Practical Tip: Preheat the oven just as you would with a conventional oven. Use an oven thermometer to monitor the internal temperature, as wood stove ovens can vary.

28. Storing and Preserving: Ensuring Long-Term Food Supply Off the Grid

Properly storing and preserving food is a crucial part of off-grid living. By using traditional and modern preservation methods such as canning, pickling, dehydration, smoking, and building a root cellar, you can extend the shelf life of your harvests and maintain a reliable food supply year-round. This chapter explores various techniques for safely preserving both produce and meat, helping you create a sustainable and efficient food storage system.

Canning: Safely Preserving Food for the Long Term

Canning is one of the most effective ways to preserve fruits, vegetables, and meats for extended periods. By sealing food in airtight jars and heating them to kill bacteria and enzymes, canning ensures that your food remains safe to eat for months or even years.

1. Understanding the Basics of Canning

There are two main methods of canning, each suited to different types of food:

Water Bath Canning

Water bath canning is used for high-acid foods, such as fruits, jams, jellies, and pickles. The high acidity helps prevent the growth of harmful bacteria, making this method safe for these types of foods.

-Best Foods for Water Bath Canning: Tomatoes, peaches, apples, berries, pickles, jams, and jellies.

-Basic Process:

1. Sterilize your jars by boiling them in water for 10 minutes.

2. Prepare your food by peeling, chopping, or cooking as needed.

3. Fill the jars, leaving about 1/2 inch of headspace.

4. Place the lids on the jars and process them in a boiling water bath for the recommended time, based on the food type and altitude.

Pressure Canning

Pressure canning is necessary for low-acid foods, such as vegetables, meats, and soups. These foods must be processed at higher temperatures to kill any potential bacteria, including Clostridium botulinum, which can cause botulism.

-Best Foods for Pressure Canning: Meats (beef, chicken, pork), vegetables (carrots, green beans, potatoes), soups, and stews.

-Basic Process:

1. Follow the same steps for preparing and filling jars as in water bath canning.

2. Use a pressure canner to process the jars at the appropriate pressure and time, ensuring that the internal temperature reaches at least 240°F (116°C).

Practical Tip: Always follow a reliable recipe for canning to ensure safety. Inaccurate times or temperatures can result in improperly preserved food, which can be dangerous.

2. Essential Equipment for Canning

Having the right tools is crucial for successful canning. Here's what you'll need:

A. Canning Jars and Lids

Use mason jars or other canning-specific jars that can withstand high temperatures. The jars come in various sizes, but pint and quart jars are the most common for home canning.

-Practical Tip: Always use new lids to ensure a proper seal. The bands (or rings) can be reused, but the flat lids should be replaced each time.

B. Water Bath and Pressure Canners

You'll need a water bath canner for high-acid foods and a pressure canner for low-acid foods. Water bath canners are large, deep pots with racks to hold the jars, while pressure canners have locking lids and pressure gauges to maintain the correct temperature.

Practical Tip: Make sure your pressure canner is properly calibrated before each use to ensure accurate pressure and safety.

Pickling: Flavorful Preservation with Vinegar and Salt

Pickling is a method of preserving food by submerging it in a brine of vinegar, salt, and spices. This method works well for preserving vegetables and fruits while adding a tangy, flavorful twist.

1. Types of Pickling: Quick vs. Fermentation

There are two main types of pickling: quick pickling and fermentation. Each offers unique benefits and flavors.

A. Quick Pickling

Quick pickling involves submerging vegetables in a vinegar-based brine and refrigerating them. This method is fast and simple but doesn't provide long-term preservation without refrigeration.

Best Vegetables for Quick Pickling: Cucumbers, carrots, radishes, onions, and peppers.

Basic Process:

1. Prepare a brine using equal parts vinegar and water, adding salt, sugar, and spices to taste.

2. Pour the hot brine over the vegetables in sterilized jars, leaving about 1/2 inch of headspace.

3. Let the jars cool to room temperature before refrigerating them.

B. Fermentation Pickling

In fermentation pickling, vegetables are submerged in a saltwater brine and left to ferment at room temperature. During fermentation, beneficial bacteria produce lactic acid, which preserves the vegetables and creates a tangy flavor.

Best Vegetables for Fermentation: Cucumbers, cabbage (for sauerkraut), carrots, and green beans.

Basic Process:

1. Dissolve salt in water to create a brine.

2. Submerge the vegetables in the brine, using a weight to keep them fully submerged.

3. Cover the jar loosely with a lid or cloth to allow gases to escape, and leave it to ferment for 1-4 weeks, depending on the temperature and desired flavor.

Practical Tip: Fermentation works best in a cool, dark place. Check the jars regularly to ensure the vegetables remain submerged, and skim off any scum that forms on the surface.

Dehydration: Drying Foods for Long-Term Storage

Dehydration removes the moisture from food, making it inhospitable for bacteria and mold, which need water to survive. Dehydrated foods are lightweight, compact, and can be stored for extended periods.

1. Methods of Dehydration

There are several ways to dehydrate food, each with its own advantages.

A. Using a Dehydrator

A food dehydrator is a specialized appliance that uses low heat and a fan to circulate air around the food, gradually removing moisture.

Best Foods for Dehydration: Fruits (apples, bananas, strawberries), vegetables (tomatoes, zucchini, peppers), herbs, and meats (for jerky).

Basic Process:

1. Slice the food thinly and evenly to ensure consistent drying.

2. Arrange the slices on the dehydrator trays, leaving space between them for air circulation.

3. Set the dehydrator to the appropriate temperature (usually 125°F to 135°F for most fruits and vegetables) and dry for 4-12 hours, depending on the food.

B. Air Drying

Air drying is an older, more traditional method that doesn't require electricity. Foods are hung or spread out in a well-ventilated area to dry naturally.

Best Foods for Air Drying: Herbs, mushrooms, and small fruits like grapes (for raisins).

Basic Process:

1. Hang or spread the food in a dry, breezy area away from direct sunlight.

2. Check the food daily and rotate or rearrange as needed to ensure even drying.

3. Drying can take several days to weeks, depending on humidity levels and the food being dried.

Practical Tip: Use cheesecloth or mesh covers to protect air-drying foods from insects.

2. Storing Dehydrated Foods

Once the food is dried, it's essential to store it correctly to maintain its shelf life.

Best Storage Practices: Store dehydrated foods in airtight containers, such as glass jars or vacuum-sealed bags, to keep out moisture and pests. Keep the containers in a cool, dark place to prolong freshness.

Shelf Life: Properly dehydrated and stored food can last anywhere from 6 months to a year, depending on the type of food.

Practical Tip: Label your containers with the date of dehydration to track how long they've been stored.

Smoking: Preserving and Flavoring Meats and Fish

Smoking is a method that both preserves food and imparts a rich, smoky flavor. By slowly cooking food over smoldering wood chips, you can create long-lasting and flavorful results.

1. Types of Smoking: Hot vs. Cold Smoking

There are two main types of smoking: hot smoking and cold smoking. Both methods have their uses, depending on the type of food and desired preservation.

A. Hot Smoking

Hot smoking cooks and preserves food simultaneously by exposing it to smoke at temperatures between 150°F and 250°F (65°C to 121°C). This method is ideal for meats and fish that will be eaten within a few days or frozen for longer storage.

Best Foods for Hot Smoking: Fish (salmon, trout), poultry, pork, and beef.

Basic Process:

1. Prepare the food by brining or seasoning it with your desired flavors.
2. Place the food in the smoker and maintain the temperature between 150°F and 250°F.
3. Smoke the food for 4-12 hours, depending on the size and type of food.

B. Cold Smoking

Cold smoking uses smoke at lower temperatures (below 90°F or 32°C) to flavor food without cooking it. This method is ideal for preserving food like cheese, bacon, or ham and requires curing the food beforehand.

Best Foods for Cold Smoking: Cheese, cured meats, and fish.

Basic Process:

1. Cure the food with salt to remove moisture and inhibit bacteria growth.
2. Cold smoke the food for 12-48 hours, depending on the food and the intensity of smoke desired.

Practical Tip: Cold smoking should only be done in cool, dry weather to prevent spoilage during the process.

Root Cellars: Natural Refrigeration for Vegetables

A root cellar is a traditional storage solution that uses the natural coolness and humidity of the earth to preserve root vegetables, fruits, and other perishable items without electricity.

1. Building a Root Cellar: Key Considerations

Building a root cellar requires planning and knowledge of your land. The goal is to create a cool, dark, and humid environment that remains relatively stable year-round.

A. Choosing the Right Location

The best location for a root cellar is underground or partially underground, where the temperature stays between 32°F and 40°F (0°C to 4°C)

Best Locations: Hillsides, basements, or areas where the ground is naturally cool.

Practical Tip: Avoid areas with poor drainage, as excess moisture can cause food to rot.

B. Designing the Root Cellar

The design of your root cellar should allow for proper ventilation and humidity control.

-Ventilation: Install vents or pipes to allow fresh air to circulate, preventing mold and spoilage.

-Shelving: Use wooden shelves or bins to store produce off the ground, where it can be kept organized and easily accessed.

2. Storing Vegetables in a Root Cellar

Different vegetables require different conditions for optimal storage. Here are the best practices for common root cellar vegetables:

Potatoes

Potatoes store best in dark, cool environments with high humidity. Keep them in bins or on shelves and cover them with burlap or straw to block light.

-Storage Time: Up to 6 months.

-Practical Tip: Keep potatoes away from apples, as apples release ethylene gas, which can cause potatoes to sprout.

Carrots and Beets

Carrots and beets can be stored in sand or sawdust to maintain humidity and prevent wilting. Bury them in layers, leaving some space between each vegetable.

-Storage Time: 4-6 months.

-Practical Tip: Leave a small amount of the greens attached to prevent moisture loss.

Apples and Pears

Fruits like apples and pears should be stored in shallow trays or crates. They need good ventilation and should be checked regularly for any signs of spoilage.

-Storage Time: 3-5 months.

-Practical Tip: Keep apples separate from other fruits and vegetables to prevent them from accelerating ripening.

Conclusion: Storing and Preserving for Long-Term Self-Sufficiency

By mastering methods like canning, pickling, dehydration, smoking, and building a root cellar, you can extend the life of your food supply and create a sustainable, year-round system for feeding your household. Each method offers unique advantages, from long-term preservation to enhancing the flavor of your food, making them indispensable tools for off-grid living.

29. Low-Tech Solutions for Off-Grid Living

One of the core elements of off-grid living is minimizing reliance on complex, high-tech solutions. However, the right low-tech alternatives can help you stay productive and maintain your independence while reducing energy consumption.

1. Hand Tools: Reliable and Energy-Free

Hand tools are indispensable for off-grid living, providing an energy-free alternative to power tools. Whether you're building, gardening, or maintaining your homestead, investing in a quality set of hand tools ensures that you can tackle essential tasks without electricity.

Essential Hand Tools for Off-Grid Living

-Hand Saws: For cutting wood or building projects, a sharp hand saw is a must-have tool. Choose a crosscut saw for general purposes and a rip saw for cutting along the grain.

-Axes and Hatchets: These are essential for chopping firewood and performing basic land clearing. A splitting axe for firewood and a hatchet for smaller tasks should be part of your toolkit.

-Shovels and Hoes: For gardening and land management, durables hovels and hoes are invaluable. Opt for tools with strong wooden or fiberglass handles for longevity.

-Manual Drills (Brace and Bit): A brace and bit drill allows you to make precise holes in wood without electricity, perfect for carpentry or simple construction projects.

Practical Tip: Keep your hand tools sharp and well-maintained to ensure they perform efficiently. Regular sharpening and oiling can extend their life and make tasks easier.

2. Manual Kitchen Appliances: Simplifying Off-Grid Cooking

Off-grid cooking can be greatly simplified with manual kitchen appliances, which eliminate the need for electricity while still allowing you to prepare meals efficiently.

Hand-Crank Kitchen Tools

-Hand-Crank Grain Mills: If you're growing your own wheat or corn, a hand-crank grain mill is a valuable tool for making flour or cornmeal. These mills are easy to operate and durable.

-Manual Coffee Grinders: For off-grid coffee lovers, a manual coffee grinder lets you enjoy fresh coffee grounds without relying on an electric grinder.

-Butter Churns: If you're raising livestock for milk, a manual butter churn is a simple and traditional way to make butter without electricity.

Practical Tip: Look for sturdy, metal construction in manual kitchen tools to ensure they last over time. Plastic components may break down with frequent use.

Simple Power Solutions: Solar-Powered Devices

In addition to manual tools, solar-powered devices provide a sustainable way to power essential items off-grid. These devices range from lighting solutions to communication tools.

Solar-Powered Lighting

Solar-powered lanterns and string lights are an excellent way to illuminate your off-grid home without relying on the grid. They recharge during the day and provide hours of light after sunset.

-Best Options: Look for rechargeable lanterns that can store energy throughout the day and provide bright, adjustable light at night.

Solar Chargers for Small Electronics

For charging small electronics like cell phones, radios, or rechargeable batteries, solar chargers are an indispensable tool. Choose a high-efficiency charger that can handle multiple devices at once for maximum utility.

Best Options: Foldable solar chargers are lightweight, portable, and easy to set up. They're ideal for camping, traveling, or remote living where access to power is limited.

30. Communication Systems: Staying Connected Off the Grid

When living off-grid, staying connected to the outside world is important for safety, security, and communication with family, friends, or emergency services. While you may not have access to a cell network or internet, several off-grid communication systems ensure that you remain in touch.

1. HAM Radios: Long-Range Communication for Off-Grid Living

HAM radios are one of the most reliable long-range communication options available to off-grid homesteads. They allow you to communicate across vast distances, even in areas where cell phone coverage is nonexistent.

How HAM Radios Work

HAM radios, also known as amateur radios, operate on specific frequency bands, allowing for long-distance communication over shortwave frequencies. These radios are especially useful in emergency situations when conventional communication systems fail.

Best Uses: Use HAM radios to communicate with nearby off-grid communities, stay updated on weather reports, or reach emergency services in remote areas.

Practical Tip: Obtain a HAM radio license to operate legally and access more frequencies. There are many online resources and local clubs that offer licensing exams and study materials.

Choosing the Right HAM Radio Setup

When selecting a HAM radio for your off-grid communication system, consider the following options:

Handheld HAM Radios: These portable radios are ideal for short-distance communication, such as within your homestead or local area.

Base Station Radios: For longer-range communication, abase station setup with a larger antenna provides the ability to reach others over hundreds of miles.

2. Satellite Phones: Reliable Communication Anywhere

When cell networks are unreliable or unavailable, satellite phones provide a direct link to the outside world. These phones connect to satellites orbiting the Earth, ensuring coverage even in the most remote locations.

Advantages of Satellite Phones

-Global Coverage: Satellite phones provide global coverage, making them ideal for emergencies or when traveling in remote areas.

-Durability: Most satellite phones are built to withstand harsh environments, with rugged exteriors that are waterproof, dustproof, and shock-resistant.

Practical Tip: While satellite phones are more expensive than other communication options, they offer unparalleled reliability in remote locations, making them a valuable investment for off-grid living.

3. Walkie-Talkies: Short-Range Communication for Everyday Use

Walkie-talkies (also known as two-way radios) are an essential tool for communicating with family members or neighbors over short distances. These radios are easy to use and require no external infrastructure like cell towers or internet.

Best Uses for Walkie-Talkies

Walkie-talkies are ideal for quick, short-range communication on your property or between nearby homes. They are especially useful for coordinating tasks on the homestead or communicating during outdoor activities like hiking or hunting.

Range: Most consumer-grade walkie-talkies have a range of 1-5 miles, depending on terrain and obstacles like trees or buildings.

Practical Tip: Choose walkie-talkies with rechargeable batteries and solar charging capabilities to ensure you always have power, even when off the grid.

31. Financial Preparedness: Budgeting for a Sustainable Off-Grid Lifestyle

Transitioning to an off-grid lifestyle requires careful financial planning to ensure that your expenses remain manageable and that you can sustain your way of life for the long term.Financial preparedness means creating a budget, managing resources wisely, and anticipating potential costs.

1. Creating a Realistic Off-Grid Budget

An off-grid lifestyle can be cost-effective, but it's important to plan for upfront investments and ongoing maintenance costs. Here's how to create a realistic budget for off-grid living:

Initial Setup Costs

When moving off the grid, you'll need to budget for the following major expenses:

-Land Purchase: Depending on the location, purchasing land for your homestead may be one of your largest investments.

-Energy Systems: Solar panels, wind turbines, and battery storage systems can be expensive upfront but save you money on energy costs in the long term.

-Water and Waste Systems: If your property lacks access to municipal water or sewage systems, you'll need to install wells, rainwater harvesting systems, or composting toilets.

Practical Tip: Plan to spend the majority of your initial budget on critical infrastructure like water, power, and shelter. These are non-negotiable and ensure your homestead is functional from day one.

Ongoing Maintenance and Costs

Even off-grid living requires ongoing expenses to maintain your property and systems. These costs may include:

-System Maintenance: Solar panels, batteries, wind turbines, and water systems all require periodic maintenance and replacement parts.

-Food and Supplies: While you may grow a portion of your food, you'll still need to purchase some items or seeds, tools, and equipment.

-Fuel Costs: If you use backup generators or wood stoves, you'll need to budget for fuel like propane, diesel, or firewood.

Practical Tip: Set aside an emergency fund to cover unexpected repairs or equipment failures. Off-grid systems can be expensive to fix, and being prepared ensures you're not left without power, water, or heat.

2. Cutting Costs and Maximizing Resources

To live sustainably off the grid, you'll need to focus on cutting costs and making the most of your available resources. Here are some strategies for reducing expenses:

Grow and Preserve Your Own Food

One of the most effective ways to cut costs is by growing and preserving your own food. A well-maintained garden or permaculture system can reduce your grocery bill significantly.

Best Practices: Invest in heirloom seeds that can be saved and replanted year after year, reducing your need to buy new seeds. Learn preservation techniques like canning, pickling, and drying to extend the shelf life of your harvest.

DIY Repairs and Upgrades

Learning to repair and upgrade your systems on your own can save you thousands of dollars in professional service fees.

Best Practices: Develop basic carpentry, plumbing, and electrical skills so that you can handle common repairs around your homestead. Invest in instructional books and videos to guide you through more complex projects.

3. Planning for Long-Term Financial Stability

While off-grid living can significantly reduce your reliance on external systems, it's important to plan forlong-term financial stability. This means managing debt, saving for the future, and anticipating potential changes in income or expenses.

Debt-Free Living

Many off-gridders aim to eliminate debt as part of their journey toward self-sufficiency. By living within your means and reducing reliance on loans or credit, you can achieve greater financial freedom.

-Practical Tip: Focus on paying off high-interest debts first and avoid taking on new loans unless absolutely necessary. Off-grid living requires flexibility, and debt can limit your ability to adapt to changing circumstances.

Building Savings

Even with careful planning, unexpected expenses will arise, whether it's a medical emergency, system failure, or natural disaster. Building a savings cushion ensures that you can cover these costs without going into debt.

Best Practices: Aim to set aside 3-6 months' worth of living expenses in an accessible savings account. This can cover emergencies or provide income if you're unable to work for a period.

32. Monetization Opportunities: Earning an Income Off the Grid

While off-grid living can reduce your expenses, finding ways to generate income is often necessary to cover ongoing costs and improve your quality of life. Fortunately, there are several monetization opportunities available for off-gridders, from selling homemade products to offering specialized services.

1. Selling Handmade Products

If you have skills in crafting, gardening, or food preservation, selling handmade products is a great way to generate extra income. This can include:

Homemade Goods

-Soap and Candles: Homemade soaps and candles are popular in local markets and online. Using natural ingredients and creative designs can help you stand out.

-Preserved Foods: Jams, jellies, pickles, and dried herbs are all in demand, particularly if you use organic or homegrown ingredients.

DIY Kits and Craft Supplies

For those with skills in woodworking, sewing, or other crafts, consider creating DIY kits that others can assemble. You could also sell raw materials like wood, wool, or seeds to fellow homesteaders.

2. Consulting and Freelance Services

Another way to earn income while living off the grid is by offering consulting or freelance services. If you have expertise in areas like permaculture, renewable energy, or sustainable living, you can monetize your knowledge through workshops, online courses, or one-on-one consulting.

Off-Grid and Sustainable Living Consulting

Many people are interested in transitioning to a more sustainable lifestyle but don't know where to start. By offering consulting services, you can share your experience and help others design their own off-grid systems.

-Practical Tip: Consider offering remote consultations via email or phone, allowing you to reach clients without needing to leave your homestead.

Writing, Blogging, and Content Creation

If you have a talent for writing or photography, you can monetize your off-grid experiences by creating content for blogs, social media, or online publications. Many people are eager to learn about off-grid living, and there's a growing market for personal stories, tutorials, and how-to guides.

-Best Practices: Focus on high-quality, practical content that offers real value to your audience. Monetize through advertising, affiliate marketing, or sponsored content.

Module I | Healthy Living, Medical Preparedness, Natural Cure, First Aid & Hygiene

33. Healthy Living and Prevention

Living off the grid brings numerous physical and mental health benefits, but it also presents unique challenges when it comes to maintaining your well-being. A lifestyle focused on self-sufficiency often requires more physical labor, a strong connection with nature, and a simpler diet. This section explores practical strategies for maintaining physical fitness, mental health, and a balanced diet while living off the grid, ensuring you remain strong and resilient in a lifestyle that demands both.

Physical Activity: Incorporating Fitness into Daily Life

Off-grid living naturally involves a great deal of physical activity, whether it's chopping wood, gardening, or building projects. However, to stay in peak condition, it's important to balance your daily tasks with a structured approach to fitness that targets different muscle groups and promotes long-term strength and endurance.

1. Daily Activities as Exercise

Many of the tasks required for off-grid living can double as exercise, allowing you to maintain fitness while getting essential work done. Here are a few ways your everyday activities contribute to physical health:

Gardening and Land Maintenance

Gardening involves bending, lifting, pulling, and digging, which engage the core, arms, legs, and back muscles. Working the land for hours a day can provide a solid cardio and strength workout.

Best Practices: Alternate between tasks to avoid overworking any one muscle group. For example, mix up weeding, hoeing, and raking to target different areas of the body.

Practical Tip: Use proper lifting techniques when handling heavy bags of soil or compost to prevent back injuries.

Firewood Chopping and Wood Stacking

Splitting firewood and stacking it for the winter is a traditional off-grid task that doubles as an intense full-body workout, especially for the arms, shoulders, and back.

Best Practices: Switch sides periodically to avoid overstraining one side of the body. Focus on using your legs to generate power during each swing of the axe, reducing strain on your back.

Practical Tip: Invest in an ergonomic axe or maul to minimize the risk of repetitive strain injuries.

Building and Construction Projects

Whether you're building a shed, repairing your home, or installing a solar panel system, off-grid projects often require heavy lifting, climbing, and working with hand tools. These activities build muscle strength, stamina, and agility.

Best Practices: Take frequent breaks during intense construction work, especially in hot weather, to avoid exhaustion and dehydration. Hydrate regularly and wear proper protective gear, such as gloves and supportive footwear.

2. Structured Fitness Routines for Off-Grid Living

While daily tasks contribute to fitness, incorporating a structured workout routine ensures that you strengthen all areas of your body and avoid potential injuries due to overuse. Focus on exercises that require no equipment or minimal equipment, such as bodyweight exercises and resistance training.

Bodyweight Exercises

Bodyweight exercises are ideal for off-grid living, as they don't require any equipment and can be done anywhere. They help build muscle strength, endurance, and flexibility.

Key Exercises:

1. Push-Ups: Strengthen the chest, shoulders, and triceps.

2. Squats: Work the legs and glutes, essential for lifting and carrying tasks.

3. Planks: Build core stability, which is crucial for maintaining good posture during heavy lifting.

4. Lunges: Target the legs and hips, improving balance and strength.

Practical Tip: Set aside 15-20 minutes a day for bodyweight exercises to supplement your daily tasks.

Resistance Bands and Improvised Weights

If you have limited space or don't want to invest in traditional gym equipment, resistance bands and improvised weights like water jugs or sandbags are excellent alternatives for strength training.

Best Practices: Use resistance bands for upper body workouts, like shoulder presses or chest expansions. Use weighted objects like sandbags for exercises such as deadlifts or rows to build back and leg strength.

Practical Tip: Create a homemade medicine ball by filling a sturdy bag with sand or rice, wrapping it tightly, and using it for exercises like squats, throws, and core rotations.

Mental Health: Cultivating Resilience and Well-Being

Maintaining good mental health is just as important as physical fitness when living off the grid. The remote nature of off-grid living can be both peaceful and isolating, so it's crucial to take steps to nurture your emotional and psychological well-being.

1. Finding Balance in Solitude

One of the challenges of off-grid living is the solitude it often brings. While many people thrive in a quiet, isolated environment, others may struggle with loneliness or lack of social interaction. Finding balance is key to mental well-being.

Embracing Solitude Positively

Spending time alone can offer an opportunity for self-reflection, creativity, and a deeper connection with nature. Use solitude as a chance to cultivate mindfulness and inner peace.

Best Practices: Incorporate daily mindfulness practices like meditation, yoga, or deep breathing exercises to stay grounded and centered. These activities help reduce stress and anxiety while improving focus and emotional regulation.

Practical Tip: Dedicate time to sit outside in nature each day, allowing yourself to be fully present in the environment. The sights, sounds, and smells of the natural world are soothing and restorative.

Maintaining Social Connections

While solitude can be beneficial, it's also important to maintain meaningful connections with others, whether through technology or in-person visits.

Practical Tip: Use solar-powered devices or low-energy communication tools like HAM radios or satellite phones to stay in touch with family, friends, or fellow off-gridders. Regular communication can prevent feelings of isolation and strengthen your support network.

2. Creating a Routine for Mental Clarity

Having a consistent daily routine helps bring structure to off-grid living and can prevent feelings of overwhelm, particularly when balancing various homesteading tasks.

The Importance of a Morning Routine

A focused morning routine sets a positive tone for the day. This might include light stretching, a healthy breakfast, and making a to-do list of essential tasks.

Best Practices: Start the day with a short walk or meditation session to get in the right mindset. Follow up by prioritizing the most critical tasks to avoid feeling scattered or unproductive.

Practical Tip: Use a journal to keep track of daily progress, accomplishments, and challenges. Reflecting on what you've achieved boosts motivation and mental clarity.

Time Management and Breaks

While off-grid living involves a lot of work, it's essential to take regular breaks to rest and recharge. Avoid burnout by alternating between physically demanding tasks and more relaxing activities.

Best Practices: Work in focused blocks of time, such as 45 minutes of work followed by a 15-minute break. Use breaks to drink water, stretch, or simply enjoy your surroundings.

Practical Tip: Incorporate leisure activities like reading, crafting, or playing music into your routine to maintain a balanced lifestyle.

Maintaining a Healthy Diet Off the Grid

Eating a nutritious diet is fundamental to staying healthy and energized, particularly when you're living off-grid and engaging in physically demanding activities. Ensuring you consume a balanced diet, rich in essential nutrients, is key to long-term health and well-being.

1. Growing and Harvesting Nutrient-Rich Foods

Off-grid living often involves growing much of your own food. A diet that includes fresh, homegrown produce is not only healthier but also more sustainable. Here are the essential steps for maintaining a balanced diet off the grid.

Key Nutrient Groups

To stay healthy, your diet should include a balance of protein, carbohydrates, healthy fats, vitamins, and minerals.

- Protein Sources: Raise chickens, goats, or rabbits for meat, eggs, and milk, which are excellent sources of protein. For plant-based protein, grow beans, lentils, and peas.

- Carbohydrate Sources: Root vegetables like potatoes, sweet potatoes, and carrots provide healthy carbohydrates, while grains like wheat, oats, and quinoa offer additional fiber and energy.

- Fruits and Vegetables: Growing a variety of fruits and vegetables ensures you get a range of vitamins and minerals. Focus on nutrient-dense produce like leafy greens, tomatoes, peppers, and berries.

Practical Tip: Incorporate companion planting and permaculture techniques to maximize your garden's yield and nutritional diversity.

Preserving Food for Year-Round Nutrition

Since off-grid living requires self-sufficiency, it's important to preserve your harvests for use during the winter or times when fresh produce is scarce.

Best Practices: Use canning, dehydration, and fermentation to preserve fruits, vegetables, and meats. Canned tomatoes, dried fruits, and fermented vegetables like sauerkraut are excellent sources of vitamins and probiotics.

Practical Tip: Build a root cellar to store root vegetables and hardy crops like squash for long-term use. This natural refrigeration method extends the life of your produce without electricity.

2. Supplementing Your Diet with Foraging and Hunting

In addition to growing food, foraging and hunting can supplement your diet with wild edibles and game. This not only provides a diverse array of nutrients but also deepens your connection with the natural environment.

Foraging for Wild Edibles

Learn to identify edible plants, nuts, mushrooms, and berries in your region. These wild foods are rich in vitamins and antioxidants, making them a healthy addition to your diet.

- Common Wild Edibles: Dandelion greens, wild garlic, blackberries, and wild mushrooms are common foraged foods that offer a wealth of nutrients.

- Practical Tip: Always forage with caution and ensure you have positively identified any wild plant before consuming it. Some wild plants can be toxic if misidentified.

Hunting for Protein

Hunting provides high-quality protein that can be preserved for long-term use. Deer, rabbits, and wild game birds are common sources of meat for off-grid communities.

Best Practices: Use smoking or dehydration to preserve meat after hunting, ensuring a long-lasting food supply.

34. Essential Medications

When living off-grid, medical emergencies can become more challenging due to limited access to healthcare professionals or pharmacies. Having a well-stocked supply of essential medications ensures you're prepared to manage common illnesses and injuries without immediate access to a doctor.

Over-the-Counter Medications: The Essentials

Over-the-counter (OTC) medications are your first line of defense for treating minor ailments at home. Below is a list of must-have medications for common issues like pain relief, fever, digestive problems, and allergies.

Pain Relief and Anti-Inflammatories

Pain relief is essential for managing headaches, muscle soreness, or injuries. Anti-inflammatory medications can help reduce swelling and discomfort.

- Ibuprofen (Advil, Motrin): Effective for reducing pain, inflammation, and fever. It's a must-have for treating injuries, muscle strains, and general pain relief.

- Acetaminophen (Tylenol, also known as Paracetamol): A great alternative for pain relief and fever reduction, particularly if you're sensitive to ibuprofen or have stomach issues.

- Aspirin: Beyond pain relief, aspirin can be used in emergency situations to manage heart attacks, making it a vital addition to your medical kit.

Practical Tip: Keep medications in a cool, dry place to maintain their efficacy over time. Regularly check expiration dates and replace outdated products.

Cold and Flu Medications

Respiratory infections, colds, and flu are common ailments, and having the right medications on hand helps manage symptoms like coughing, congestion, and fever.

- Cough Syrup: Choose an OTC cough syrup with dextromethorphan for suppressing a dry, unproductive cough.

- Decongestants (Pseudoephedrine, Phenylephrine): These help reduce nasal congestion and sinus pressure, making it easier to breathe.

- Antihistamines (Diphenhydramine, Loratadine): Useful for relieving allergy symptoms, sneezing, and runny noses, and can also aid in sleeping when you're ill.

Practical Tip: If you live in a remote or cold climate, stock up on cold and flu medications, especially during the winter months when respiratory illnesses are more common.

Digestive Medications

Digestive issues are common, especially when adapting to a new diet or dealing with stress. Keep medications on hand for indigestion, diarrhea, and constipation.

- Antacids (Tums, Rolaids): Quickly neutralize stomach acid to relieve indigestion or heartburn.

- Loperamide (Imodium): Essential for treating diarrhea, especially during times when access to clean water may be limited.

- Bismuth Subsalicylate (Pepto-Bismol): Useful for treating nausea, upset stomach, and diarrhea.

Practical Tip: Diarrhea can lead to dehydration quickly, so it's also helpful to stock up on oral rehydration salts or make your own solution with water, salt, and sugar.

Allergy and Skin Care Medications

Skin irritations, insect bites, and allergies are common issues when living close to nature. Having antihistamines and topical treatments is essential for treating allergic reactions, rashes, and minor skin conditions.

- Hydrocortisone Cream: Reduces itching, swelling, and redness caused by insect bites, allergic reactions, or skin irritations.

- Antihistamine Tablets (Diphenhydramine, Cetirizine): Useful for managing mild allergic reactions or seasonal allergies.

- Calamine Lotion: Provides relief from itching caused by poison ivy, sunburn, or other irritants.

Practical Tip: If you live in an area with high insect activity, consider keeping insect repellents and antihistamine cream in your medical kit to prevent and treat bites.

Wound Care and Antiseptics

Minor injuries like cuts, scrapes, and burns are common, especially when performing physical labor. Proper wound care is crucial for preventing infection.

- Antibiotic Ointment (Neosporin, Bacitracin): Helps prevent infection in minor cuts, scrapes, and burns.

- Sterile Gauze and Bandages: Keep a variety of bandage sizes and sterile gauze pads for covering wounds.

- Hydrogen Peroxide or Isopropyl Alcohol: These antiseptics are used for cleaning wounds and preventing infection.

Practical Tip: For more severe injuries, stock butterfly bandages or steri-strips to close deeper wounds temporarily before seeking medical care.

Prescription Medications: Planning for Chronic Conditions and Emergencies

If you have a chronic health condition, it's important to plan ahead to ensure you have an adequate supply of prescription medications. Here's how to manage prescription needs in an off-grid setting.

Stocking Up on Prescription Medications

For those with conditions like diabetes, hypertension, or thyroid disorders, maintaining a consistent supply of prescription medications is essential.

- Best Practices: Work with your healthcare provider to get an extended supply of medications or use mail-order pharmacies that offer larger quantities. Keep prescriptions in their original packaging, which includes important dosing information and expiration dates.

Emergency Medications

Certain prescription medications are essential in emergencies, particularly for allergic reactions, infections, or injuries.

- Epinephrine (EpiPen): If you or someone in your household has severe allergies, keep EpiPens on hand for emergency use during anaphylactic reactions.

- Antibiotics: In certain cases, your doctor may provide a supply of antibiotics for use in case of severe infections. However, it's important to use antibiotics responsibly to prevent antibiotic resistance.

Practical Tip: Have a medication log to track your supplies and expiration dates, and regularly review your stock to ensure you're prepared for any situation.

35. Natural Cures: Growing and Using Herbal Remedies

Living off-grid offers the perfect opportunity to explore natural cures and herbal remedies for common ailments. Many plants and herbs have medicinal properties that can be used to treat everything from digestive issues to minor infections. Here's how you can incorporate herbal medicine into your off-grid healthcare system.

Medicinal Plants You Can Grow Off the Grid

There are many easy-to-grow plants with medicinal properties that thrive in a variety of climates. Whether you grow them in a garden or in pots on your porch, these plants offer a sustainable source of natural medicine.

Aloe Vera

Aloe Vera is known for its soothing properties, especially for treating burns, cuts, and skin irritations. It's a hardy plant that can thrive both indoors and outdoors with minimal care.

- Uses: Apply the gel from fresh aloe vera leaves directly to burns, sunburns, or irritated skin. Aloe is also used for digestive health when taken internally in small doses.

- Practical Tip: Keep aloe vera near your kitchen or outdoor cooking area for easy access in case of burns.

Peppermint

Peppermint is a fast-growing herb that can be used to treat digestive problems, headaches, and respiratory issues. It can be grown in pots or garden beds and thrives in both full sun and partial shade.

- Uses: Brew fresh or dried peppermint leaves into tea to relieve indigestion, gas, or nausea. Inhaled as steam, peppermint can help open the sinuses and relieve congestion.

- Practical Tip: Dry and store peppermint leaves for year-round use. When dried, peppermint can retain its medicinal properties for several months.

Chamomile

Chamomile is well-known for its calming effects and is often used to relieve stress, promote sleep, and reduce inflammation. This herb grows best in well-drained soil with full sunlight.

- Uses: Chamomile tea is commonly used to treat insomnia, anxiety, and indigestion. It can also be applied topically as a poultice to reduce inflammation in skin irritations or minor wounds.

- Practical Tip: Harvest chamomile flowers when they are fully open, and dry them to create a long-lasting supply for tea or topical use.

Echinacea

Echinacea is commonly used to boost the immune system and shorten the duration of colds and flu. This herb is relatively easy to grow in most climates and is resistant to drought once established.

- Uses: Echinacea tea or tinctures can be used to stimulate the immune system, helping to fend off colds, flu, and other infections. It can also be used topically to treat wounds and prevent infection.

- Practical Tip: Harvest the roots, leaves, and flowers for medicinal use. Echinacea tinctures made from the roots are particularly potent and can be stored for long-term use.

Lavender

Lavender is a fragrant herb known for its calming effects, as well as its ability to reduce inflammation and treat minor skin irritations. It thrives in well-drained soil and full sunlight.

- Uses: Lavender oil can be applied to the skin to treat burns, insect bites, or rashes. Its aroma also promotes relaxation and relieves stress, making it useful in aromatherapy.

- Practical Tip: Dry lavender flowers to use in sachets for a calming aroma or to make lavender-infused oils for topical use.

Preparing and Using Herbal Remedies

Once you've grown your medicinal herbs, it's important to know how to properly prepare them for use. Here are the most common ways to prepare and use herbal remedies.

Herbal Teas and Infusions

Herbal teas are one of the easiest ways to take medicinal herbs internally. They are made by steeping fresh or dried herbs in hot water.

- How to Prepare: Use 1-2 teaspoons of dried herbs or a small handful of fresh herbs per cup of water. Steep for 10-15 minutes, then strain and drink.

- Best Herbs for Teas: Peppermint, chamomile, echinacea, and ginger are all excellent herbs for making medicinal teas.

Practical Tip: For stronger medicinal effects, you can make an infusion by steeping the herbs in hot water for several hours or overnight. This method extracts more active compounds from the plant.

Tinctures and Extracts

Tinctures are concentrated liquid extracts made by soaking herbs in alcohol or vinegar for several weeks. Tinctures are highly potent and can be stored for years without losing their effectiveness.

- How to Make a Tincture: Fill a jar halfway with dried herbs, then cover with alcohol (vodka or brandy) or vinegar. Let the mixture sit in a cool, dark place for 4-6 weeks, shaking it daily. Afterward, strain the liquid and store it in a dark glass bottle.

- Best Herbs for Tinctures: Echinacea, valerian, and lavender are commonly used in tinctures for their potent medicinal properties.

Practical Tip: Use a dropper to take small doses of tinctures (usually 1-2 droppers full) as needed. Tinctures are particularly useful for immune support and stress relief.

Salves and Poultices

Herbal salves and poultices are used for topical application, helping to treat skin conditions, wounds, and inflammation.

- Salves: Made by infusing herbs in oil and then thickening with beeswax, salves are ideal for treating burns, cuts, or dry skin.

- Poultices: A poultice is made by crushing fresh herbs and applying them directly to the skin. They are used to reduce swelling, treat wounds, or soothe insect bites.

Practical Tip: Keep a supply of beeswax and olive oil on hand to make homemade salves. These can be stored for long-term use and applied as needed.

36. First Aid Preparation: Equipping Yourself for Medical Emergencies

A well-equipped first aid kit is essential for managing injuries, infections, and other medical emergencies when living off-grid. Additionally, having a basic understanding of first aid techniques is just as important as having the supplies themselves. Below, we'll cover the core items that should be in your first aid kit, as well as the essential first aid skills you'll need to stay prepared for anything.

Basic First Aid Supplies: What to Include in Your Kit

Your first aid kit should be tailored to the unique risks of off-grid living, such as cuts, burns, bites, or allergic reactions. Here's a comprehensive list of basic items that should be in every off-grid first aid kit:

Wound Care Supplies

Wounds are one of the most common injuries in off-grid living, especially if you're using tools or working outdoors. Proper wound care helps prevent infections and accelerates healing.

- Sterile Gauze Pads and Rolls: Used to cover wounds and stop bleeding.

- Adhesive Bandages (Various Sizes): Essential for covering small cuts, blisters, or abrasions.

- Antiseptic Wipes or Solution: Clean wounds and prevent infections with wipes soaked in alcohol or other antiseptics.

- Hydrogen Peroxide or Saline Solution: Use to clean out dirt and debris from wounds before applying dressings.

- Antibiotic Ointment (e.g., Neosporin): Apply to wounds to prevent bacterial infections.

- Sterile Gloves: Always wear gloves when treating wounds to avoid contaminating them.

Practical Tip: Store wound care supplies in airtight containers to keep them sterile, and check expiration dates regularly to ensure they are still effective.

Burn Care

With activities like cooking over open flames or working with hot tools, burns are a risk in off-grid settings. Burn care supplies are critical for managing minor burns and preventing infection.

- Burn Gel or Aloe Vera: These products help cool the skin and relieve pain.

- Non-stick Sterile Dressings: Apply these to burns to protect the skin and prevent the dressing from sticking to the wound.

- Burn Cream (with Lidocaine): Provides pain relief for minor burns and accelerates healing.

Practical Tip: For more severe burns, cover the affected area loosely with a non-stick dressing and seek medical attention as soon as possible. In an off-grid setting, having access to emergency communication tools is crucial for these situations.

Medications and Treatments

Your first aid kit should also include basic medications for treating pain, inflammation, allergic reactions, and infections.

- Ibuprofen or Acetaminophen: These pain relievers are essential for managing aches, pains, and fever.

- Antihistamines (e.g., Diphenhydramine): Used to treat allergic reactions from insect bites or plant exposure.

- Epinephrine Auto-Injector (EpiPen): If anyone in your household is at risk for anaphylaxis, keep this life-saving medication on hand.

- Oral Rehydration Salts: Useful for treating dehydration caused by heat, illness, or diarrhea.

- Eye Wash Solution: Flush out dirt, debris, or chemicals from the eyes.

Practical Tip: Rotate medications regularly to ensure you always have non-expired supplies, and store them in a cool, dry place for maximum shelf life.

Splints and Bandages

Accidents that result in sprains, fractures, or dislocations may require splinting to immobilize the affected area.

- Elastic Bandages (Ace Wraps): Useful for wrapping sprains or securing splints in place.

- Sam Splints or Foam Splints: Lightweight, versatile splints that can be molded to stabilize broken bones or sprains.

- Triangular Bandage: Can be used as a sling for arm injuries or to secure splints.

Practical Tip: Learn how to improvise splints using materials available on your homestead, such as sticks or boards. Elastic bandages can also serve multiple purposes, making them a versatile tool in your first aid kit.

Advanced First Aid Supplies for Off-Grid Settings

In addition to basic supplies, certain advanced items are invaluable when living off the grid, where immediate medical assistance may not be available.

Hemostatic Agents

For severe bleeding that can't be controlled with pressure alone, hemostatic agents (e.g., QuikClot) are used to promote rapid clotting.

- Best Practices: Apply the hemostatic agent directly to the wound before applying a pressure dressing to control life-threatening bleeding.

Practical Tip: Make sure you understand how to use these products safely, as improper use could cause complications.

Suture Kit or Wound Closure Strips

If you're far from medical assistance, being able to close wounds is a critical skill. While suture kits are more advanced, wound closure strips (like Steri-Strips) provide a simpler alternative for closing deeper cuts.

- Best Practices: Clean the wound thoroughly before applying closure strips, and keep the area immobilized while it heals.

Practical Tip: Consider taking a wilderness first aid course to learn how to use suture kits and other advanced first aid tools effectively.

Basic First Aid Training: Essential Skills for Self-Reliance

Even with a well-stocked first aid kit, having the knowledge to properly assess and treat injuries is crucial. Here are the top first aid skills to learn before heading off the grid:

Wound Management and Infection Prevention

Knowing how to clean and dress wounds effectively can prevent infections and promote faster healing.

- Best Practices: Always clean wounds with antiseptic solutions and use sterile dressings. For deep or large wounds, use compression bandages to control bleeding and prevent shock.

CPR and Emergency Response

Cardiopulmonary resuscitation (CPR) is an essential life-saving skill that can make a critical difference in emergencies.

- Practical Tip: Take a CPR certification course, which is often available through organizations like the Red Cross. This skill is invaluable for anyone, especially when living in remote areas where help may not arrive quickly.

Fracture and Sprain Treatment

Sprains, strains, and fractures are common in an off-grid lifestyle due to the physical labor involved. Knowing how to immobilize an injury until proper care is available is vital.

- Best Practices: Immobilize the affected area with a splint, keep the limb elevated, and apply cold compresses to reduce swelling.

37. Hygiene: Staying Clean and Disease-Free Off the Grid

Maintaining personal hygiene is crucial for preventing illness and infection, especially when living off-grid without access to running water or modern sanitation facilities. In this section, we'll explore practical hygiene strategies that ensure you stay clean and healthy in an off-grid environment.

Water Conservation and Hygiene Without Running Water

When access to water is limited, conserving and using it efficiently is essential for maintaining hygiene. Here are some techniques to stay clean without a constant supply of running water.

Dry Bathing

Dry bathing with wet wipes, alcohol wipes, or homemade cleansing cloths is an excellent alternative to traditional bathing when water is scarce.

- Best Practices: Focus on cleaning essential areas—face, underarms, and groin—daily with wipes or a damp cloth. Use soap and water on hands and feet regularly, as they are most likely to come into contact with contaminants.

Practical Tip: Keep a supply of biodegradable wipes in your hygiene kit. These can be composted after use, making them an environmentally friendly option.

Sponge Baths

When a small amount of water is available, a sponge bath is a simple way to stay clean.

- Best Practices: Heat a small basin of water over your stove, and use a washcloth to scrub your body with soap and warm water. Focus on key areas where bacteria tend to accumulate.

Practical Tip: Set up a designated area for sponge bathing with privacy screens if living with others, ensuring comfort and hygiene for everyone.

Maintaining Oral Hygiene Off the Grid

Oral hygiene is just as important as bodily hygiene, especially since dental care may be difficult to access in remote areas. Neglecting oral health can lead to infections, tooth decay, or worse. Here's how to maintain a healthy mouth off-grid.

Basic Oral Hygiene Supplies

At a minimum, you'll need a toothbrush, toothpaste, and dental floss to keep your teeth and gums healthy.

- Best Practices: Brush your teeth twice daily and floss once per day to remove food particles and plaque. Choose fluoride toothpaste, as it helps prevent cavities and strengthens tooth enamel.

Practical Tip: If you run out of commercial toothpaste, you can make a simple alternative using baking soda mixed with a little water. Baking soda helps neutralize acids and remove plaque.

Handwashing and Sanitization Without Running Water

Keeping hands clean is one of the best ways to prevent the spread of disease, especially after using the bathroom or handling food. Without running water, you'll need to get creative with your handwashing solutions.

Portable Handwashing Stations

A simple portable handwashing station can be made with a large water container, a spigot, and a bucket for catching used water.

- Best Practices: Fill the container with clean water, add a small amount of biodegradable soap, and use the spigot to wash your hands over the bucket. Dispose of the gray water by irrigating plants or filtering it for reuse.

Practical Tip: Keep hand sanitizer in your hygiene kit for situations when you can't set up a full handwashing station. Choose alcohol-based hand sanitizers with at least 60% alcohol for maximum effectiveness.

Managing Waste: Sanitation and Disease Prevention

Properly managing human waste is one of the most important factors in preventing disease in off-grid environments. Without access to a traditional sewer system, you'll need alternative solutions for handling waste safely.

Composting Toilets

Composting toilets are an eco-friendly and sanitary option for managing waste without water. They convert human waste into compost through natural processes, reducing the risk of contamination.

- Best Practices: Use sawdust, wood chips, or other carbon-rich materials to cover waste after each use. This helps control odors and speeds up the composting process. Regularly rotate the compost to ensure proper decomposition.

Practical Tip: Make sure to keep your composting toilet well-ventilated to avoid unpleasant odors and ensure that the composting process works efficiently.

Greywater Recycling

Recycling greywater (the relatively clean wastewater from sinks, showers, or baths) is a sustainable way to conserve water and reduce waste.

- Best Practices: Set up a simple greywater recycling system to divert water from your sinks or showers to irrigate your garden or flush your composting toilet.

Practical Tip: Avoid using harsh chemicals in your cleaning products or soaps, as they can contaminate your greywater and harm your plants.

Module J | Off-Grid Education & Skills Acquisition

Living off-grid requires a unique set of skills that differ from urban or suburban life. The entire family — adults and children alike — needs to be equipped with practical knowledge that fosters independence, resourcefulness, and self-sufficiency. This section covers the core skills necessary for thriving off the grid, including fire-starting, gardening, and basic first aid, as well as DIY and handcrafting skills such as carpentry, blacksmithing, and knitting.

38. Homeschooling and Child Education

One of the most rewarding aspects of off-grid living is the ability to shape your children's education in alignment with your values and lifestyle. Homeschooling provides the flexibility to focus on practical skills, critical thinking, and real-world problem-solving, alongside traditional academic subjects. In this section, we'll explore strategies for structuring an effective off-grid education, the subjects and skills that are essential, and how to integrate learning into daily life.

Tailoring Homeschooling for Off-Grid Living

Homeschooling in an off-grid environment offers a unique opportunity to design a curriculum that goes beyond traditional academics. By incorporating life skills, survival techniques, and environmental awareness, you can create a holistic education that prepares your children for both modern challenges and the demands of off-grid living.

Designing a Flexible Curriculum

Off-grid homeschooling allows for flexibility in both subjects and teaching methods, which means you can customize your children's education to suit their interests, learning styles, and future goals.

Core Academic Subjects

While practical skills are crucial for off-grid living, maintaining a strong foundation in core academic subjects ensures that your children are well-rounded and prepared for any future opportunities, whether they decide to pursue higher education or stay within the off-grid lifestyle.

- Mathematics: Emphasize problem-solving and applied math, such as using math in construction projects, gardening, or budgeting.

- Language Arts: Focus on reading comprehension, writing, and critical thinking. Encourage children to keep a journal of their off-grid experiences, or help them start a family newsletter.

- Science: Tailor science lessons to the natural world around you. Biology, ecology, and earth sciences are especially relevant, as they help children understand the environment they live in. Use your homestead as a living laboratory for learning.

Practical Tip: Create project-based learning opportunities that combine multiple subjects. For example, have your children design a small garden bed using mathematical calculations for area, and then research the best plants for your climate (science).

Practical Life Skills

One of the greatest benefits of homeschooling off the grid is the ability to focus on life skills that are crucial for self-reliance. These skills are often neglected in traditional schooling but are invaluable for off-grid living.

- Gardening and Food Preservation: Teach children how to plant, care for, and harvest crops, as well as how to preserve food through canning, drying, and pickling.

- Basic Carpentry: Introduce children to tools and building techniques by starting with small projects like birdhouses or raised garden beds.

- First Aid: Make first aid training part of their curriculum, focusing on both minor injuries (cuts, sprains) and more serious situations, like choking or fractures.

Practical Tip: Let your children take the lead on certain homestead projects, such as planning and planting a garden or building simple structures. This empowers them to apply their skills in a real-world context.

Creating a Daily Homeschool Routine

Establishing a daily routine is key to ensuring your children stay on track academically while also engaging in the many hands-on learning opportunities that off-grid life offers.

Balancing Academics and Practical Learning

One of the advantages of off-grid homeschooling is that you can blend academic learning with practical experience. Here's how to create a balanced schedule:

- Morning Academics: Start the day with structured learning in core subjects like math, reading, and science. These subjects require focus and concentration, making the morning an ideal time to tackle them.

- Afternoon Practical Skills: Use the afternoons for hands-on activities like gardening, cooking, or outdoor exploration. Practical learning reinforces academic concepts while keeping children engaged.

Practical Tip: Rotate subjects and activities throughout the week to keep things fresh. For example, dedicate Monday mornings to math and Wednesday afternoons to building projects or nature walks.

Incorporating Seasonal Learning

Off-grid living is deeply connected to the seasons, and homeschooling should reflect this natural rhythm. Tailor your lessons to the seasonal changes on your homestead.

- Spring and Summer: Focus on outdoor learning—gardening, nature walks, animal care, and construction projects. This is the time to teach about the life cycle of plants, soil health, and sustainable farming practices.

- Fall and Winter: Move indoors for lessons on food preservation, crafts, and DIY projects. Teach children to can and preserve the summer's harvest or make homemade candles and soap.

Practical Tip: Keep a seasonal learning journal to track the natural changes on your homestead. Have your children record observations, sketch plants and animals, and note how their activities change with the seasons.

Focusing on Key Subjects for Off-Grid Education

While traditional subjects like math, science, and literature are important, homeschooling off the grid provides a unique opportunity to teach essential off-grid skills and critical thinking. Here are the subjects and skills that should be prioritized in an off-grid homeschooling curriculum.

Environmental Science: Understanding Nature

A deep understanding of environmental science is critical for living in harmony with nature. Use your homestead as a laboratory to teach concepts like ecology, renewable energy, and conservation.

Ecology and Ecosystems

Teach your children about the ecosystem they live in, including the plants, animals, and insects that inhabit your homestead. Understanding how all the elements of an ecosystem interact is key to sustainable living.

Best Practices: Have children identify the native species in your area and research their roles in the ecosystem. Encourage them to keep a nature journal where they record observations and discoveries.

Practical Tip: Take regular nature walks and encourage children to document what they see. Use their observations to discuss topics like food chains, pollination, and biodiversity.

Renewable Energy and Sustainability

Off-grid living often requires using renewable energy sources like solar, wind, or hydro power. Teaching children about these systems is both practical and educational.

Best Practices: Include hands-on projects that involve setting up small solar panels, building wind turbines, or creating water-powered generators.

Sustainability Projects: Teach children how to calculate their energy consumption and explore ways to reduce their carbon footprint through energy-efficient practices.

Practical Tip: Have older children research and present a project on how renewable energy systems work. Let them design their own miniature solar system or wind turbine to demonstrate their understanding.

Practical Mathematics: Applied Problem-Solving

While traditional math instruction is important, focus on applied mathematics that can be used in daily off-grid tasks. This helps children see the real-world value of math and how it applies to homesteading.

Building and Construction Math

Construction projects on your homestead require measuring, estimating, and planning—all skills that can be taught through applied math.

Best Practices: Have children measure the dimensions of a raised garden bed, calculate the area needed for planting, or figure out how much lumber is required for a building project.

Practical Tip: Create math challenges related to homesteading tasks, such as estimating how much water is needed to irrigate a garden or calculating the slope for a drainage system.

Budgeting and Resource Management

Teach children how to budget and manage the resources on your homestead, such as food, energy, and water. Learning to manage resources responsibly is a vital life skill.

Best Practices: Involve children in family budgeting discussions. Teach them how to track expenses for homesteading projects or how to plan meals based on the food you've grown and preserved.

Practical Tip: Use real-world examples to teach fractions, percentages, and basic economics. For instance, show children how to calculate savings from growing your own vegetables versus buying them from a store.

Integrating Learning into Daily Life

One of the greatest benefits of homeschooling off-grid is the ability to integrate learning into everyday activities. This creates a more natural and engaging education for children and ensures they are constantly building both academic and life skills.

Learning Through Homestead Projects

Everyday tasks on your homestead provide countless learning opportunities. Whether it's tending to animals, building structures, or preserving food, you can use these projects as hands-on lessons.

Animal Care and Biology

Raising livestock offers a perfect opportunity to teach children about animal biology and veterinary care.

Best Practices: Have your children care for chickens, goats, or rabbits, and teach them the basics of animal anatomy, feeding, and health care.

Practical Tip: Use animal care to teach responsibility and accountability. Assign daily chores like feeding and cleaning, and discuss the importance of ethical and humane treatment of animals.

Cooking and Chemistry

Cooking is an essential life skill, and it's also a way to teach children chemistry and mathematics.

Best Practices: Let children participate in meal planning and preparation. Discuss the chemical reactions that happen during cooking, such as yeast rising in bread or the effects of heat on food.

Practical Tip: Introduce children to preservation methods such as canning or fermenting. These processes teach practical chemistry while also contributing to the family's food supply.

Building a Family Learning Environment

Creating a home that encourages learning is essential for homeschooling success. This includes setting aside spaces for studying, reading, and exploring hobbies or projects.

Designating Learning Areas

Designate specific areas in your home or outdoor spaces for different types of learning. This helps children associate certain areas with specific activities and creates a routine for them to follow.

Indoor Learning Spaces

Set up a small study area indoors where children can focus on academic work like reading, writing, and math.

Best Practices: Keep this area well-organized with materials like notebooks, pencils, and reference books easily accessible.

Practical Tip: Use a chalkboard or whiteboard for teaching lessons or working through problems together as a family.

Outdoor Learning Areas

Designate a part of your homestead as an outdoor classroom, where children can engage in activities like gardening, building projects, or nature observation.

Best Practices: Set up a workbench for outdoor projects or a nature table where children can collect and display interesting rocks, leaves, and insects they find.

Practical Tip: Use outdoor spaces for hands-on experiments or physical education, such as obstacle courses or games that teach coordination and teamwork.

Fostering Independence and Lifelong Learning

Homeschooling off the grid is an opportunity to create a rich, well-rounded education that goes beyond traditional academics. By focusing on practical skills, critical thinking, and self-reliance, you prepare your children for the unique challenges of off-grid living while also equipping them with the knowledge and adaptability they need for life.

39. Essential Off-Grid Skills for Adults and Children

Fire-Starting: Mastering a Vital Survival Skill

Starting a fire is a fundamental skill for off-grid living, especially when it comes to heating, cooking, and even signaling for help in emergencies. Both adults and children should learn how to start and manage a fire safely.

Basic Fire-Starting Methods

Mastering a few different fire-starting techniques is essential for being prepared in various conditions. Teaching children these skills early builds confidence and preparedness.

- Friction-Based Fire Starting: Learn to create a fire using friction-based tools like a bow drill or hand drill. These methods take practice but are excellent for emergencies.

- Flint and Steel: Keep a flint and steel kit in your gear for reliable fire-starting in any weather condition.

- Fire Starters and Tinder: Create homemade fire starters using natural materials such as dried leaves, pinecones, or lint. Knowing how to find or prepare dry tinder is crucial for getting a fire going in damp conditions.

Practical Tip: Teach children the importance of fire safety—how to build a fire in a safe area, keep it controlled, and extinguish it properly.

Maintaining a Fire

Once a fire is started, maintaining it for a long period is another critical skill.

- Building a Sustainable Fire: Teach the different types of fire setups, such as the teepee, log cabin, or lean-to fire, depending on your needs.

- Choosing the Right Fuel: Understanding the types of wood or fuel available is key. Use softwoods like pine for quick burning, and hardwoods like oak for longer-lasting fires.

- Managing the Fire Safely: Educate children and adults alike on how to control a fire, adjust its size, and put it out safely by dousing it with water or smothering it with dirt.

Practical Tip: Store dry kindling and firewood in a weatherproof location to ensure a steady supply, especially in wet conditions.

Gardening: Cultivating Food for Self-Sufficiency

Growing your own food is a cornerstone of off-grid living, and it's important for every family member to know how to plant, maintain, and harvest crops. Gardening teaches responsibility, problem-solving, and the basics of food production.

Planting and Soil Preparation

Start with learning how to prepare the soil for gardening. Soil health directly affects crop yields, and understanding how to improve and maintain it is essential for a successful garden.

- Testing Soil: Learn how to test and amend your soil with compost or natural fertilizers.

- Planting: Learn the optimal times for planting, spacing, and depth for seeds or seedlings.

- Composting: Teach the family how to create a composting system to produce nutrient-rich soil for the garden.

Practical Tip: Encourage children to plant their own section of the garden, choosing fast-growing crops like radishes or beans to maintain their interest.

Maintaining and Harvesting Crops

Once your plants are established, maintaining their health through regular watering, weeding, and pest control is key.

- Weeding: Teach children and adults to recognize weeds and remove them before they compete with crops for nutrients and space.

- Watering Techniques: Set up a drip irrigation system or use rainwater harvesting techniques to minimize water usage while keeping crops hydrated.

- Harvesting: Understanding when and how to harvest crops ensures you get the best yield and prevents spoilage. Teach the family how to identify ripeness and proper harvesting techniques.

Practical Tip: Use surplus harvests for preserving (canning, pickling, or drying) to ensure food availability year-round.

Basic First Aid

Knowing how to treat minor injuries and health conditions off-grid is vital, particularly when professional medical help might not be immediately available. Everyone in the household should learn basic first aid techniques, including how to treat cuts, burns, sprains, and other common injuries.

First Aid Kit Essentials

A well-stocked first aid kit is the foundation of off-grid healthcare. Teach children and adults how to use the items in the kit for common injuries.

- Wound Care: Understanding how to clean and dress cuts and scrapes is important to avoid infection.

- Burn Treatment: Learn how to treat burns using aloe vera, cool compresses, and burn ointments.

- Splinting: Knowing how to immobilize an injured limb with a makeshift splint from sticks or other materials can prevent further injury.

Practical Tip: Practice first aid skills with family members so everyone feels confident in managing minor emergencies.

Treating Illnesses Off the Grid

In addition to injury care, being able to manage illnesses like colds, fevers, or infections is crucial when off the grid.

- Hydration: Teach children the importance of staying hydrated, especially when sick, and how to use oral rehydration solutions.

- Natural Remedies: Incorporate herbal medicine into your first aid plan by growing herbs like echinacea for immune support or peppermint for digestive issues.

Practical Tip: Regularly rotate and replace first aid kit supplies to ensure everything is fresh and functional when needed.

40. DIY and Handcrafting Skills

Building, fixing, and creating things by hand is essential to off-grid living. These DIY skills ensure that you're less reliant on external resources and better prepared to tackle daily challenges. Whether it's basic carpentry, blacksmithing, or knitting, these skills improve your self-reliance and can even become a source of income.

Carpentry: Building Structures and Furniture

Carpentry is a valuable skill in off-grid living, allowing you to build or repair shelters, furniture, and tools. Here's what every off-gridder should know:

Basic Carpentry Tools and Techniques

- Essential Tools: Start with basic hand tools like saws, hammers, chisels, and planes. These are versatile and don't require electricity.

- Building Simple Structures: Learn how to create shelters, outbuildings, or even fences with basic carpentry skills.

- Furniture Making: Crafting tables, benches, or shelves from reclaimed wood can save money and add personal touches to your home.

Practical Tip: Work on small carpentry projects with children, such as birdhouses or simple wooden toys, to teach them these essential skills.

Blacksmithing

Blacksmithing is a traditional skill that can help you make and repair metal tools, hardware, and even knives. It requires a forge, basic equipment, and practice, but is invaluable in an off-grid setting.

Blacksmithing Basics

- Setting Up a Forge: Learn how to build a simple coal or gas-powered forge and get the right tools for basic blacksmithing.

- Forging Simple Tools: Start with basic items like nails, hooks, or simple garden tools before moving on to more complex projects like blades or farm implements.

Practical Tip: If blacksmithing feels too advanced, consider learning metalworking skills for repairing and sharpening tools rather than forging new ones.

Knitting and Textile Skills: Clothing Repair and Production

Knitting, sewing, and textile repair are critical skills for maintaining clothing and making items like blankets or rugs from available resources.

Basic Knitting and Sewing Techniques

- Knitting: Learn basic stitches like the garter stitch and purl stitch to create hats, scarves, or blankets.

- Sewing: Being able to mend or repair clothing ensures that your family's clothing lasts longer. Learn how to patch tears, sew buttons, and hem clothing.

Practical Tip: Keep a simple sewing kit with needles, thread, and fabric patches on hand for quick repairs.

41. Skill Sharing and Apprenticeship

One of the greatest advantages of living off the grid is the opportunity to build a community of like-minded individuals who share knowledge and skills. By participating in skill-sharing networks or becoming an apprentice, you can expand your skill set while contributing to your community.

Building a Local Off-Grid Community

Living off-grid doesn't have to mean isolation. Building a local network of fellow off-gridders can help you share knowledge, trade goods, and offer mutual support in emergencies.

Skill Sharing with Neighbors

Create a skill-sharing system with your neighbors where everyone contributes based on their strengths.

- Carpentry and Construction: One family might excel in building, while another has expertise in gardening or herbal medicine.

- Barter and Trade: Trade goods like preserved foods, homemade tools, or handcrafted items in exchange for services or materials.

Practical Tip: Establish a regular community meet-up or workshop where you can learn new skills from others and teach what you know.

Apprenticeship: Learning by Doing

An apprenticeship is one of the best ways to learn a new trade, especially when it comes to skills like blacksmithing, carpentry, or farming.

Finding Mentors

Seek out experienced off-gridders who are willing to teach you the skills you need. Offer to help with their projects in exchange for learning new techniques.

Learning Hands-On: Apprenticeships allow you to practice as you learn, building confidence in your new skills.

Practical Tip: If no one nearby is available, seek online courses or videos that focus on the off-grid skills you're interested in, and practice them on your own projects.

42. Online and Offline Learning Resources

While off-grid living might suggest limited access to the internet or other technologies, many off-gridders balance periods of being fully offline with occasional internet access through solar-powered devices, satellite internet, or visits to nearby towns. This opens up a wealth of resources for learning new skills and improving your homesteading experience.

Online Learning Resources: Harnessing Technology for Off-Grid Education

Even though living off the grid often involves a reduction in technology use, online learning remains a valuable tool for acquiring new knowledge and skills. Here are some of the best options available for off-gridders.

eBooks and Online Libraries

Digital libraries and eBooks are an excellent resource for homesteaders looking to expand their knowledge. You can download books on topics like renewable energy, gardening, construction, and more.

- Free Online Libraries: Websites like Project Gutenberg and Open Library offer thousands of free eBooks on a wide range of topics, many of which are relevant to off-grid living.

- Kindle Books: Kindle eBooks are a portable way to store a vast collection of manuals and guides on your tablet or e-reader, allowing you to access them even when offline.

Practical Tip: Download books and guides while you have access to the internet, so they're available for reference when you're offline.

Online Courses and Tutorials

Many websites offer online courses in everything from carpentry and food preservation to herbal medicine and renewable energy systems. These courses often provide detailed, step-by-step instructions, making them ideal for self-paced learning.

- Udemy and Coursera: Both of these platforms offer courses on a variety of subjects that can be directly applied to off-grid living, such as permaculture design, basic carpentry, and solar power installation.

- YouTube: Many off-grid experts share tutorials on topics like DIY construction, woodworking, or gardening. Videos offer a visual learning experience, which can be especially helpful for complex tasks.

Practical Tip: Save videos to your device when you have internet access, so you can watch them offline when needed.

Remote Learning Communities

Off-gridders can benefit from remote learning communities, where individuals share experiences, advice, and skills. These communities often offer the opportunity to join discussion forums, attend webinars, or participate in virtual meetups.

- Facebook Groups and Reddit: Join off-grid and homesteading communities where members share their knowledge, post solutions to common problems, and discuss everything from DIY projects to survival tips.

- Skillshare: This platform allows users to take classes on a wide range of topics, and you can connect with instructors and fellow learners to expand your skills in areas like self-sufficiency or handcrafting.

Practical Tip: Participate in online discussions or Q&A sessions to get personalized advice from experienced homesteaders.

Offline Learning Resources: Traditional Knowledge for Modern Off-Grid Living

When internet access is limited or unavailable, offline learning resources become invaluable. Books, manuals, and local mentors are essential tools for continuous education, ensuring you have the knowledge to solve problems and grow your skills.

Essential Books for Off-Grid Living

Building a well-rounded library of reference books is one of the best ways to ensure you're prepared for anything in an off-grid environment. Here are some must-have categories:

- Gardening and Permaculture: Books like "The Permaculture Handbook" by Peter Bane and "Gaia's Garden" by Toby Hemenway offer deep insights into sustainable farming and gardening techniques.

- Construction and Carpentry: Manuals like "The Complete Manual of Woodworking" by Albert Jackson provide essential instructions for DIY building projects.

- Survival and Preparedness: Titles like "The Survival Medicine Handbook" by Joseph Alton provide step-by-step guidance for emergency medical situations when professional help is unavailable.

Practical Tip: Organize your books by topic and keep them in a central location on your homestead for easy access during projects.

Physical Manuals and Guides

Hard copies of instruction manuals and field guides are essential, particularly for skills that require in-depth understanding, such as plumbing, electrical systems, or off-grid power generation.

- Renewable Energy Manuals: Books like "Solar Power for Beginners" or "Off the Grid: A Simple Guide to Solar, Wind, and Geothermal Energy" provide detailed information on setting up and maintaining renewable energy systems.

- Herbal Medicine Guides: Keep comprehensive guides on medicinal herbs, such as "The Herbal Medicine-Maker's Handbook" by James Green, to learn how to grow, harvest, and use herbs for treating common ailments.

Practical Tip: Annotate your manuals with personal notes and observations as you implement their advice. This will make it easier to reference key points when you need them.

Local Learning Resources and Mentors

In addition to books and manuals, learning from local experts and mentors is invaluable. Communities often have experienced individuals who are willing to share their knowledge, especially on topics like farming, construction, and food preservation.

- Farmers and Homesteaders: Local farmers or fellow off-gridders can offer firsthand knowledge about the challenges and opportunities specific to your region.

- Workshops and Meetups: Attend local events or workshops that focus on skills like beekeeping, herbal medicine, or livestock care. Even in rural areas, these events can often be found through local bulletin boards or online.

Practical Tip: Build a network of local mentors and attend regular meetups to continue learning and improving your skills. Bartering services or skills in exchange for lessons can be a great way to establish these relationships.

43. Critical Thinking and Problem-Solving: Adaptability for Off-Grid Challenges

Critical thinking and problem-solving skills are key to successfully navigating the unpredictable challenges of off-grid living. Without the convenience of immediate professional assistance or readily available supplies, developing the ability to think critically and improvise is essential. This section will guide you in teaching critical thinking, problem-solving, and adaptability, all of which are crucial for thriving off the grid.

Developing Problem-Solving Skills Through Real-World Challenges

In an off-grid environment, problem-solving often revolves around addressing immediate, practical concerns. From mechanical failures to crop management, every challenge is an opportunity to sharpen your problem-solving skills.

Mechanical and System Failures

Whether it's your solar power system or water filtration unit, mechanical failures can occur without warning. Developing troubleshooting skills is essential for diagnosing and repairing these issues.

- Identify the Problem: Break down the issue by isolating individual components. For example, if your solar panel isn't generating power, check the charge controller, inverter, and battery connections systematically.

- Test Solutions: Work through potential fixes step by step. Try replacing fuses, adjusting settings, or cleaning connectors.

Practical Tip: Keep a troubleshooting journal where you record common issues, fixes, and outcomes. Over time, this log will become a valuable resource for diagnosing future problems more efficiently.

Improvising Tools and Materials

Off-grid living often requires you to improvise when you don't have the ideal tools or materials available. Learning to adapt and create solutions with what you have on hand is a critical off-grid skill.

- DIY Tools: If you lack a specific tool, figure out how to craft a makeshift version. For instance, a piece of wood can serve as a temporary lever, or a simple tarp and poles can create an effective shelter.

- Repurposing Materials: Old materials, like used lumber or metal, can be repurposed for various projects. Turn leftover wood into garden beds, or fashion scrap metal into small parts for repairs.

Practical Tip: Keep a stockpile of scrap materials on hand and regularly practice using them for creative solutions to everyday problems.

Fostering Critical Thinking Skills in Children

Teaching children how to approach challenges with critical thinking is just as important as teaching them practical skills. The earlier they learn to assess situations, evaluate options, and make decisions, the more self-sufficient and adaptable they will become.

Encourage Curiosity and Exploration

Encouraging curiosity in children helps them develop the skills to analyze and question the world around them. Use real-world tasks on the homestead to foster critical thinking.

- Exploration: Let children experiment with tasks like building a small shelter, fixing a broken tool, or setting up an irrigation system.

- Asking Questions: Encourage children to ask why certain things happen. For example, why do some plants thrive while others struggle in your garden? Have them investigate factors like soil quality, water levels, and sunlight.

Practical Tip: Create small challenges for children where they have to solve problems or puzzles using available materials, such as setting up a simple water filtration system using natural resources.

Teaching Decision-Making

Good decision-making requires the ability to weigh options, evaluate risks, and make informed choices. Teach children to assess situations and make thoughtful decisions based on the information at hand.

- Risk Assessment: When faced with a task like crossing a creek or climbing a tree, discuss with children the potential risks and benefits of their actions.

- Evaluating Options: Present multiple solutions to a problem and ask children to determine which is the best approach. For example, should they use solar energy or wind power to charge a battery, depending on current weather conditions?

Practical Tip: Involve children in decision-making processes on the homestead, such as planning a garden layout or choosing the best way to repair a tool. Encourage them to consider the pros and cons of each option.

Problem-Solving Strategies for Adults: Building Resilience

In addition to teaching children, adults must also continuously refine their problem-solving abilities. Off-grid living often presents unexpected challenges, and the ability to remain calm and methodical is crucial for long-term success.

The Importance of Planning and Foresight

One of the best ways to avoid crises is through proactive planning. Anticipate potential problems before they arise and take steps to prevent them.

- Predicting Failures: Regularly inspect and maintain critical systems like water pumps, solar panels, or fencing to prevent breakdowns.

- Stockpiling Essentials: Ensure you have essential supplies like extra food, water, tools, and repair materials on hand for emergencies.

Practical Tip: Create a monthly checklist of systems and resources to inspect and maintain. This proactive approach will reduce the likelihood of unexpected breakdowns or shortages.

Staying Calm Under Pressure

When emergencies do arise, staying calm and working through the problem systematically is key to finding a solution.

- Breathing and Focus: In a crisis, take a few deep breaths to calm your mind. Then, break the problem into smaller, more manageable parts.

- Working in Teams: If possible, involve others in problem-solving efforts. Different perspectives can help generate creative solutions and distribute the workload.

Practical Tip: Involve your entire family in mock drills for various emergency scenarios, such as a power outage, flooding, or a medical emergency. Practicing these situations builds confidence and preparedness for real events.

Module K | Off-Grid Entertainment and Mental Wellbeing

44. Board Games and Group Activities: Fun Without Electricity

One of the simplest ways to bond with family and friends while living off the grid is through board games and other group activities that don't require electricity. These activities help you pass the time, reduce stress, and foster connections, making them an essential part of off-grid life.

Classic Board Games: Timeless Fun for All Ages

Board games offer a timeless way to have fun and engage the whole family. Many classic games can be played repeatedly without ever getting old, and they require nothing more than the game pieces and a few players.

Chess and Checkers

Chess and checkers are excellent choices for off-grid entertainment because they stimulate strategic thinking and patience. These games also provide an opportunity for multiple players to improve their skills over time.

Best Practices: Set up a designated area in your home or outdoor space where these games can be played regularly, encouraging family members to challenge each other in a friendly competition.

Practical Tip: If you don't have a chess or checkers set, consider crafting your own using natural materials like wood, clay, or even painted rocks for the pieces.

Card Games

A simple deck of cards can provide endless entertainment, with games ranging from solitaire for individuals to more complex group games like Rummy, Poker, and Crazy Eights.

Best Practices: Rotate through different card games to keep things fresh and fun. Teach children and adults alike how to shuffle and deal, creating a family tradition around weekly card game nights.

Practical Tip: Cards are portable and easy to store, making them perfect for camping trips or outdoor gatherings.

Trivia and Word Games

For those who love a mental challenge, trivia and word games are excellent choices for group activities. Games like Scrabble, Pictionary, and 20 Questions can be both educational and fun.

Best Practices: Customize trivia games to your family's interests, such as creating questions around homesteading, history, or nature. This adds a learning element to the game and makes it more relevant to your off-grid lifestyle.

Practical Tip: Use a chalkboard or whiteboard to keep score, and offer small prizes like homemade treats or extra chore help for the winners.

Outdoor Group Activities: Staying Active Together

In an off-grid setting, outdoor activities offer not only entertainment but also physical exercise and opportunities for bonding.

Scavenger Hunts

Scavenger hunts are a fun way to explore your surroundings while teaching children about nature. Create a list of items to find, such as specific leaves, rocks, or flowers, and see who can collect the most in a set amount of time.

Best Practices: Tailor your scavenger hunts to the seasons, incorporating natural elements that change throughout the year, such as snow in winter or flowers in spring.

Practical Tip: Incorporate educational elements, like identifying different tree species or types of wildlife, to add value to the activity.

Capture the Flag and Other Group Games

For larger groups or families, games like Capture the Flag, Hide and Seek, or Tag are great for encouraging physical activity and teamwork. These games can be played outdoors and require no equipment, making them ideal for off-grid living.

Best Practices: Set boundaries for the playing area to keep everyone safe, especially if your homestead is surrounded by forest, water, or uneven terrain.

Practical Tip: Rotate teams and switch up the games to keep things exciting and inclusive for all ages.

45. Hobbies and Crafts: Cultivating Creativity and Practical Skills

Off-grid living provides ample time and space to develop new hobbies and engage in crafting projects that are both enjoyable and practical. From woodworking to sewing, these activities allow you to create functional items for your homestead while also cultivating a sense of accomplishment.

Woodworking: Crafting with a Purpose

Woodworking is not only a useful skill for building and repairing structures, but it's also a rewarding hobby that allows you to create everything from furniture to decorative items.

Building Simple Projects

Start with small, manageable projects such as birdhouses, garden planters, or picture frames. These require minimal tools and can be completed over a weekend.

Best Practices: Keep a set of hand tools like saws, chisels, and sandpaper readily available. Work with locally sourced or reclaimed wood to make your projects more sustainable.

Practical Tip: Involve children in basic woodworking tasks to teach them valuable skills while spending quality time together.

Advanced Woodworking Projects

As your skills progress, you can take on more complex projects like building furniture, constructing raised garden beds, or even designing small cabins or sheds.

Best Practices: Plan your projects carefully by sketching out designs and measuring twice before cutting. Having a clear plan helps you avoid mistakes and makes the process more efficient.

Practical Tip: Craft gifts for family members or trade your handmade items with neighbors to build community ties and share your talents.

Sewing and Textiles: Creating Functional and Artistic Pieces

Sewing is an essential skill for off-grid living, allowing you to repair clothing, create new garments, and craft items for your home. Beyond practicality, sewing can also be a relaxing and creative hobby.

Repairing and Repurposing Clothing

Learning to repair clothing is vital for making your wardrobe last longer. Patching holes, sewing buttons, and hemming garments are basic but essential skills.

Best Practices: Keep a sewing kit stocked with needles, thread, fabric scraps, and buttons. Regularly check your family's clothing for minor issues and repair them before they become major problems.

Practical Tip: Host a clothing repair workshop with friends or neighbors, where everyone brings their worn-out clothes and you help each other make repairs while sharing tips and techniques.

Crafting Quilts, Curtains, and Home Decor

Once you've mastered basic sewing, you can move on to creating quilts, curtains, or other home decor items. These projects not only beautify your space but also provide warmth and comfort during colder months.

Best Practices: Use fabric scraps or repurposed materials to make quilts and other textiles. This is a great way to reduce waste and create meaningful, handmade items for your home.

Practical Tip: Start a long-term quilt project where family members can contribute pieces over time, creating a collaborative heirloom that tells the story of your off-grid journey.

Gardening as a Hobby: Growing More Than Just Food

While gardening is a necessity for most off-gridders, it can also be an enjoyable hobby that offers relaxation, beauty, and a sense of accomplishment. Beyond growing vegetables, consider planting flowers, herbs, and decorative plants.

Flower Gardening

Planting a flower garden provides color and joy, attracting pollinators like bees and butterflies, which are essential for your vegetable garden's success.

Best Practices: Choose flowers that thrive in your climate and are easy to maintain, such as sunflowers, marigolds, or wildflowers. Mix in herbs like lavender or mint for added fragrance and utility.

Practical Tip: Start a cut flower garden where you can harvest fresh bouquets for your home or trade with neighbors.

Herb Gardening for Medicinal and Culinary Uses

Growing herbs such as basil, rosemary, and echinacea offers both practical benefits and the joy of watching plants grow. Herbs can be used for cooking, crafting herbal remedies, and making teas.

Best Practices: Plant herbs in pots or small raised beds for easy access. You can also dry them for use in the winter months or for creating homemade herbal remedies.

Practical Tip: Create a herb-drying station in your home where you can air-dry your herbs, ensuring you have a steady supply year-round.

46. Music and Storytelling

Music and storytelling have been used for centuries as a means of entertainment, education, and community building. In an off-grid setting, these traditions can be revived to create shared experiences without the need for modern technology.

Playing Instruments: Keeping Music Alive

Whether you play an instrument or sing, music is a powerful way to express creativity and connect with others. Instruments like guitars, banjos, and harmonicas are perfect for off-grid living because they require no electricity and can be played anywhere.

Learning and Teaching Instruments

If you or a family member plays an instrument, take the time to teach others. This fosters a shared love of music and creates opportunities for family jam sessions.

Best Practices: Schedule regular practice times or impromptu musical gatherings around a campfire or in the living room. Encourage everyone to participate, even if they're just singing along.

Practical Tip: If you're new to an instrument, take advantage of manuals or basic music theory books to learn the basics of chords, scales, and rhythm.

Storytelling: Preserving Oral Traditions and Creating New Ones

Storytelling is one of the oldest forms of entertainment and education. By sharing stories around the dinner table or campfire, you can pass down family history, lessons, and traditions while also sparking creativity and imagination.

Telling Traditional Stories

Many cultures have long traditions of folk tales and legends that offer moral lessons, humor, or history. Incorporating these stories into your family's routine keeps these traditions alive.

Best Practices: Create a storytelling night where each family member takes turns telling a story. These can be personal anecdotes, historical tales, or even fictional stories created on the spot.

Practical Tip: Encourage children to develop their storytelling skills by letting them create and perform their own stories for the family.

47. Physical Exercise - Staying Fit Off the Grid

Maintaining physical health is vital for living an active, off-grid lifestyle. Incorporating regular physical exercise into your routine keeps you strong, resilient, and prepared for the demands of homesteading.

Outdoor Physical Activities: Exercising in Nature

Outdoor activities like hiking, chopping wood, and gardening provide both exercise and practical benefits for your homestead.

Hiking and Exploring

Hiking is a great way to explore your surroundings while staying fit. It strengthens muscles, improves cardiovascular health, and connects you to nature.

Best Practices: Set aside time each week for family hikes. Choose paths that challenge different fitness levels, gradually increasing the difficulty as your endurance builds.

Practical Tip: Turn hikes into nature scavenger hunts to keep children engaged and learning while getting exercise.

Yoga and Body-Weight Exercises: Staying Strong Indoors

For days when outdoor exercise isn't possible, body-weight exercises like push-ups, squats, and planks are effective for building strength without the need for equipment.

Simple Yoga Routines

Yoga improves flexibility, strength, and balance—all of which are important for preventing injury and maintaining physical health on the homestead.

Best Practices: Incorporate daily yoga routines that focus on core strength, balance, and stretching. These routines can be short and adapted to different skill levels.

Practical Tip: Create a designated exercise space in your home or yard where you can practice yoga, stretch, or do body-weight exercises without distractions.

48. Cultivating Mental Clarity and Emotional Balance

Mindfulness and meditation are powerful practices that can help you manage stress, stay centered, and maintain emotional balance while living off the grid. Whether you're dealing with the challenges of homesteading or simply trying to find peace in a remote environment, practicing mindfulness can make your off-grid experience more fulfilling.

The Benefits of Mindfulness in an Off-Grid Setting

Mindfulness is the practice of staying present and fully engaged in the moment. In an off-grid environment, where life is often slower and more connected to nature, mindfulness can help you savor each moment and reduce stress.

Enhancing Focus and Patience

Living off-grid often requires a high level of focus and patience, whether you're maintaining a garden, building a structure, or caring for animals. Practicing mindfulness can help you stay focused on the task at hand and cultivate patience, even when faced with challenges.

Best Practices: Take a few minutes each morning to set your intentions for the day. Focus on being present in each task, whether it's chopping wood, planting seeds, or preparing meals.

Practical Tip: Use the natural rhythms of your environment to guide your mindfulness practice—listen to the sound of the wind in the trees, the flow of a nearby stream, or the crackle of the fire as you practice staying present in the moment.

Managing Stress and Anxiety

Even in the peaceful setting of off-grid living, stress can arise from unexpected challenges like harsh weather, equipment failures, or isolation. Practicing mindfulness helps you respond to these challenges calmly and rationally.

Best Practices: When you feel stressed, pause and take three slow, deep breaths. Focus on your breathing to help calm your mind and body before addressing the issue at hand.

Practical Tip: Set aside time each day for a short mindfulness meditation session. Find a quiet spot outdoors and focus on your breathing, gently bringing your mind back to the present whenever it wanders.

Meditation Techniques for Off-Grid Living

Meditation is a practice that helps you cultivate a sense of inner peace and mental clarity. It's an essential tool for maintaining emotional balance, especially when you live in a remote environment with fewer social outlets.

Breath Awareness Meditation

Breath awareness meditation is a simple but powerful technique that can be practiced anywhere, making it perfect for off-grid living.

Best Practices: Find a comfortable, quiet spot—either indoors or outdoors. Sit or lie down in a relaxed position, close your eyes, and focus on your breath. Pay attention to each inhale and exhale, and whenever your mind wanders, gently bring your focus back to your breath.

Practical Tip: Practice breath awareness for just 5–10 minutes each day. Over time, you'll find that this simple practice helps reduce stress and improve your overall sense of wellbeing.

Nature Meditation

Living off the grid offers a unique opportunity to incorporate nature into your meditation practice. Nature meditation involves connecting with the natural world around you in a mindful way.

Best Practices: Sit or walk quietly in a natural setting, such as a forest, meadow, or by a body of water. Focus your attention on the sounds, sights, and smells around you—the rustling leaves, the songs of birds, or the feel of the earth beneath your feet.

Practical Tip: Create a meditation garden or quiet outdoor space where you can retreat for mindfulness practice. Surround yourself with plants, rocks, or water features that enhance the meditative atmosphere.

Gratitude Meditation

Gratitude meditation is a practice that helps you focus on the positive aspects of your life, cultivating a sense of appreciation for the simple pleasures of off-grid living.

Best Practices: At the end of each day, take a few moments to reflect on what you're grateful for—whether it's the beauty of nature, the warmth of your home, or the satisfaction of a hard day's work.

Practical Tip: Keep a gratitude journal where you write down three things you're grateful for each day. This practice can help shift your mindset to a more positive and appreciative perspective, even in the face of challenges.

49. Managing Social Isolation: Staying Connected in Remote Environments

One of the challenges of off-grid living is the potential for social isolation, especially if you live in a remote area with few neighbors or limited access to towns. While isolation can be a welcome break from the fast pace of modern life, it's important to balance it with meaningful social interactions to avoid loneliness or cabin fever.

Recognizing the Signs of Isolation and Cabin Fever

Social isolation can lead to feelings of loneliness, boredom, and restlessness. It's important to recognize these signs early and take steps to address them before they impact your mental health.

Symptoms of Cabin Fever

Cabin fever is a term used to describe the feelings of restlessness and irritability that can arise from being confined in one place for an extended period. Symptoms can include:

- Boredom or Lethargy: Feeling unmotivated or uninterested in daily tasks.

- Irritability or Anxiety: Becoming easily frustrated or anxious about small issues.

- Restlessness: A constant feeling of wanting to get out or do something different.

Practical Tip: If you notice these symptoms in yourself or family members, take immediate steps to break up the routine. Change your environment by spending more time outdoors or engaging in a new hobby.

Overcoming Loneliness

Loneliness can be particularly challenging in remote off-grid settings, but there are strategies for preventing and overcoming it.

Best Practices: Make an effort to stay connected with loved ones through letters, satellite phones, or occasional visits to town. Regular communication with family and friends helps maintain your emotional ties, even when you're physically distant.

Practical Tip: Schedule weekly or monthly check-ins with friends and family members, either in person or through remote communication methods. Having these connections to look forward to can help prevent feelings of isolation.

Finding Solitude Without Loneliness

While too much isolation can be harmful, learning to enjoy solitude without feeling lonely is one of the greatest benefits of off-grid living. Solitude offers a chance for deep reflection, creativity, and personal growth.

Embracing Quiet Time

Solitude allows you to focus on personal goals, engage in creative pursuits, or simply enjoy the quiet beauty of nature. Learning to embrace this quiet time can be deeply fulfilling.

Best Practices: Use periods of solitude to engage in creative projects, such as writing, painting, or building. Set aside time each day for activities that allow you to express yourself and recharge mentally.

Practical Tip: Create a personal sanctuary in your home or outdoors—a quiet space where you can retreat for reading, meditation, or contemplation.

50. Building Community Connections: Creating Social Networks in Off-Grid Living

Even though you may be living off the grid, building connections with others is still important for both practical and emotional support. Finding or creating a community of like-minded individuals can enhance your off-grid experience, provide valuable resources, and combat isolation.

Finding Like-Minded Off-Grid Families and Neighbors

Living off the grid doesn't necessarily mean living in isolation. Many off-grid communities or individuals are eager to share knowledge, offer support, and connect with others who share their values.

Attending Local Markets and Events

Many rural and off-grid communities have farmers' markets, craft fairs, or community events that bring people together. These events are great opportunities to meet neighbors, share experiences, and exchange ideas.

Best Practices: Attend these events regularly, not only to stock up on supplies but also to build relationships with fellow off-gridders. Offer to trade goods or services, such as fresh produce, handcrafts, or homesteading skills.

Practical Tip: Join or form a local barter network where families can trade goods, services, or labor. This fosters a sense of community and helps you connect with others who are living similarly.

Joining Online Off-Grid Communities

If physical distance makes it difficult to connect with nearby families, consider joining online communities where off-gridders share knowledge, experiences, and advice.

Best Practices: Participate in forums, social media groups, or video chats with other off-gridders. These virtual communities provide valuable support and camaraderie, even if you're geographically isolated.

Practical Tip: Host or join a virtual meet-up with other off-grid families to discuss specific topics like gardening, renewable energy, or homeschooling. These gatherings allow you to share insights and build connections despite physical distance.

Creating a Strong Local Support Network

Building a local support network is essential for off-grid living, especially in emergencies or when you need help with large projects. A strong local network provides practical assistance and fosters a sense of belonging.

Helping Neighbors

One of the best ways to build strong community ties is through acts of service. Offer to help neighbors with their projects, whether it's building a structure, harvesting crops, or troubleshooting a solar power system.

Best Practices: Trade skills and resources with your neighbors. If you have a particular expertise, such as carpentry or herbal medicine, offer to help in exchange for something they're skilled in, such as mechanics or livestock care.

Practical Tip: Organize seasonal work parties, where families come together to help with major tasks like building, planting, or preserving food. These events strengthen community bonds and provide mutual support.

51. Reading and Personal Education: Cultivating Knowledge and Creativity

One of the best ways to continue learning and stay mentally engaged in an off-grid setting is through reading. Books offer an endless source of knowledge, inspiration, and entertainment, making them an essential part of your off-grid lifestyle.

Creating a Reading Nook: A Space for Quiet Reflection

Designating a space in your home for reading and personal education provides a quiet retreat for learning, reflection, and relaxation.

Setting Up a Comfortable Space

Choose a cozy corner of your home or cabin where you can create a reading nook. Include comfortable seating, soft lighting, and shelves for your favorite books.

Best Practices: If you have access to natural light, set up your reading space near a window. Incorporate blankets, pillows, or natural elements like plants or wood for a soothing, peaceful atmosphere.

Practical Tip: Keep a rotating selection of books in your reading nook, from reference manuals to novels and memoirs. This allows you to switch between educational and recreational reading as your mood dictates.

Building a Personal Library

A personal library is a valuable resource in an off-grid setting, especially if you're seeking to become more self-sufficient. Stock your shelves with books on homesteading, survival skills, DIY projects, and more.

Best Practices: Organize your books by category, such as gardening, herbal medicine, animal care, and renewable energy. Having quick access to reference materials will make it easier to troubleshoot issues and continue learning.

Practical Tip: Swap books with friends or neighbors to keep your library fresh. This not only adds variety to your collection but also strengthens community ties through shared resources.

Personal Education: Expanding Your Knowledge in Off-Grid Living

Reading isn't just for entertainment; it's also a crucial part of self-education in an off-grid lifestyle. Continuing to learn new skills and ideas keeps you mentally sharp and better prepared for the challenges of homesteading.

Choosing Practical Reading Material

In addition to fiction or personal interest books, focus on practical guides and manuals that expand your homesteading skills. Look for books on topics like:

- Permaculture and Gardening: Learn how to grow food sustainably and manage your land.
- DIY and Carpentry: Build your own furniture, structures, and tools.
- Renewable Energy: Gain a deeper understanding of solar, wind, and water power systems.

Practical Tip: Set personal education goals, such as learning a new skill every season. Read books that support these goals and apply your learning through hands-on projects on your homestead.

Module L | Off-Grid Transportation & Mobility

Living off the grid presents unique challenges when it comes to transportation and mobility. Without the convenience of modern infrastructure, it's important to develop strategies for independent travel and vehicle maintenance. This part explores the best methods for staying mobile in an off-grid environment, including alternative transportation, vehicle repair, fuel independence, and emergency preparedness.

52. Alternative Transportation Methods

In an off-grid setting, relying solely on motorized vehicles may not always be practical or sustainable. Exploring alternative transportation methods allows you to minimize reliance on fossil fuels, reduce costs, and stay mobile in a variety of conditions.

Bicycles and Electric Bikes

Bicycles and electric bikes (e-bikes) are ideal for off-grid transportation, especially for short trips to town, nearby farms, or neighbors. These options are eco-friendly, easy to maintain, and don't require fuel.

Choosing the Right Bike

When selecting a bike for off-grid living, focus on durability, versatility, and ease of maintenance.

Best Practices: Opt for sturdy mountain bikes or all-terrain bikes that can handle rough paths and uneven terrain. E-bikes can be solar-charged, giving you extra mobility with minimal energy input.

Practical Tip: Keep a repair kit on hand with tools like tire levers, a patch kit, and a pump to handle minor issues on the go.

Electric Bike Maintenance

E-bikes offer additional convenience, but they require regular maintenance to ensure long-term usability.

Best Practices: Regularly check the battery, brakes, and tire pressure. Store your bike in a dry, protected area to prevent weather damage, and ensure your battery is charged with solar power to maximize efficiency.

Practical Tip: Install a small solar charging station for your e-bike, ensuring you have sustainable power even in remote areas.

Animal-Powered Transport

In more rural or remote settings, animal-powered transport is a time-tested method of mobility. Horses, donkeys, and mules can provide reliable transportation as well as assistance with farming tasks.

Choosing the Right Animal

When deciding on an animal for transport, consider your terrain, climate, and workload.

Best Practices: Donkeys and mules are excellent for mountainous or rough terrain, while horses offer speed and endurance. Ensure your animal is trained for riding and carrying supplies or pulling carts.

Practical Tip: Set up a rotational grazing system to keep your animals healthy and well-fed, reducing the need for supplemental feed.

Caring for Work Animals

Animal-powered transportation requires consistent care, including feeding, grooming, and veterinary checkups.

Best Practices: Keep a basic veterinary kit on hand for treating minor injuries and maintaining good hoof health. Rotate your animals between work and rest to prevent overexertion.

Practical Tip: Learn basic farrier skills so you can trim your animal's hooves and fit horseshoes yourself, reducing dependency on outside help.

Boats and Watercraft: Mobility for Water-Based Living

If your off-grid home is near a lake, river, or coastline, a small boat or canoe can be an excellent method of transport for fishing, gathering supplies, or traveling to nearby communities.

Choosing the Right Watercraft

The type of watercraft you choose depends on the local waterways and your transportation needs.

Best Practices: Canoes and kayaks are perfect for small rivers and lakes, while rowboats or small motorized boats can handle larger bodies of water. Ensure you have a life jacket for each passenger and practice basic water safety.

Practical Tip: Use solar panels to charge small motors or batteries for electric watercraft, reducing fuel consumption.

Electric Vehicles: Sustainable Off-Grid Mobility

For those with access to solar or wind energy, electric vehicles (EVs) provide an eco-friendly transportation option. With proper infrastructure, EVs can be a viable long-distance transportation method for off-gridders.

Solar Charging for EVs

Setting up a solar charging station for your electric vehicle allows you to remain independent from fuel sources.

Best Practices: Position solar panels in areas with optimal sunlight exposure and ensure they are connected to an inverter and battery system large enough to charge your vehicle's battery.

Practical Tip: Install a portable solar charger that can travel with you, enabling you to charge your vehicle on the go during long journeys.

53. Vehicle Maintenance

Keeping your vehicles in good working condition is critical, especially in remote areas where access to professional mechanics may be limited. Learning basic vehicle maintenance and setting up a small auto shop on your homestead ensures you can handle repairs when needed.

Basic Repairs and Maintenance

Whether you rely on trucks, ATVs, or electric vehicles, knowing how to perform basic repairs and maintenance is essential for off-grid living.

Regular Maintenance Routines

Performing routine maintenance helps prevent costly breakdowns and keeps your vehicle running smoothly.

Best Practices: Change the oil, filters, and fluids regularly. Check tire pressure, brake systems, and battery health as part of your regular maintenance routine.

Practical Tip: Keep a maintenance log for each vehicle to track repairs, oil changes, and other key information. This helps you stay on top of regular upkeep.

Handling Common Issues

Some of the most common vehicle problems can be fixed with basic tools and know-how.

Best Practices: Learn how to fix a flat tire, replace a battery, and repair a fuel line. Keep a set of essential spare parts like extra fuses, spark plugs, and brake pads.

Practical Tip: Watch online tutorials on basic vehicle repair before attempting larger projects to ensure you're comfortable with the process.

Setting Up an Off-Grid Auto Shop

Having a designated space for vehicle repairs and maintenance is a game-changer when living off the grid. Setting up a small auto shop with essential tools and parts allows you to handle most repairs yourself.

Essential Tools for Vehicle Maintenance

Stock your shop with the basic tools needed for common vehicle repairs and upgrades.

Best Practices: Include wrenches, sockets, jack stands, and screwdrivers in your tool kit. A hydraulic jack and tire iron are essential for lifting vehicles and changing tires.

Practical Tip: Keep a manual for each vehicle in your shop so you can reference it during repairs.

Storing Parts and Supplies

In addition to tools, your shop should have a stockpile of essential parts and consumables like oil, brake fluid, and coolant.

Best Practices: Store parts in a dry, organized space to prevent corrosion or damage. Label shelves for easy access to frequently used items like filters, hoses, and belts.

Practical Tip: Barter with neighbors or local mechanics for spare parts if you're unable to find them yourself.

54. Fuel Independence

To maintain mobility, off-gridders often need to ensure fuel independence. This section covers how to produce your own biodiesel, store fuel safely, and manage emergency fuel supplies.

Biodiesel Production: Powering Your Vehicles Sustainably

One of the most effective ways to maintain fuel independence is by producing your own biodiesel. This renewable fuel can be made from used cooking oil, animal fats, or plant oils.

Biodiesel Production Process

The process of making biodiesel involves mixing oils with an alcohol (such as methanol) and a catalyst (usually lye) to create a fuel that can be used in diesel engines.

Best Practices: Set up a small biodiesel production station with safety equipment, including gloves and goggles. Ensure you have enough storage containers for both the raw materials and the finished biodiesel.

Practical Tip: Keep a step-by-step guide to the biodiesel production process handy to ensure you follow the right ratios and procedures.

Fuel Storage: Safely Storing Gasoline, Diesel, and Propane

For those who rely on gasoline, diesel, or propane for transportation or backup power, safe storage is essential to prevent fires, explosions, or fuel degradation.

Long-Term Fuel Storage Solutions

Fuel can degrade over time, so it's important to store it properly to maximize shelf life.

Best Practices: Store fuel in airtight, UV-resistant containers in a cool, dry location. Use fuel stabilizers to extend the life of stored gasoline or diesel.

Practical Tip: Set up a rotational system where you use the oldest stored fuel first to ensure nothing goes to waste.

55. Emergency Vehicles

In an off-grid environment, having the right emergency vehicles can be crucial during natural disasters or other urgent situations. Vehicles like ATVs, boats, and emergency rafts provide reliable transportation over difficult terrain.

All-Terrain Vehicles (ATVs): Navigating Rough Terrain

ATVs are highly versatile and can handle rough or uneven terrain, making them an excellent choice for emergency transportation in rural or mountainous areas.

Maintaining Your ATV

ATVs require regular maintenance, including checking the brakes, tires, and fluids.

Best Practices: Keep spare tires, spark plugs, and filters on hand to ensure your ATV is always ready for use in an emergency.

Practical Tip: Install a small trailer to your ATV to carry essential supplies, tools, or first aid kits.

Boats and Emergency Rafts: Preparedness for Floods and Water-Based Emergencies

For those living near bodies of water or in flood-prone areas, having an emergency boat or raft on hand is essential for evacuation and transportation.

Choosing the Right Emergency Watercraft

Inflatable rafts and small motorboats are useful for navigating floods, while canoes and kayaks offer quieter, more efficient travel on calm waters.

Best Practices: Keep emergency flotation devices and life jackets with your boat or raft. Make sure the boat is easily accessible in case of flooding.

Practical Tip: Practice loading your family and essential supplies into the boat during dry conditions to ensure you can evacuate quickly and efficiently in an emergency.

56. Self-Repair and Essential Tools

To ensure your transportation methods remain reliable, you need to develop a solid understanding of vehicle repair and maintenance. This section outlines the most important tools to keep on hand for vehicle upkeep and repair in an off-grid setting.

Essential Tools for Off-Grid Repair

Having the right tools allows you to perform basic repairs on your vehicles, reducing downtime and ensuring you stay mobile.

Basic Tool Kit for Vehicle Repair

Include tools such as wrenches, sockets, screwdrivers, pliers, and a jack in your kit. These tools will allow you to handle most basic vehicle repairs, from changing tires to fixing leaks.

Best Practices: Organize your tools in a portable tool chest so they're easy to transport in case of breakdowns away from home.

Practical Tip: Learn to improvise tools from natural materials or repurpose household items for repairs in case you're missing specific tools.

Stocking Spare Parts

Maintaining a supply of essential spare parts ensures you can repair your vehicle without waiting for external support, especially in remote areas.

What to Keep on Hand

Stock parts like filters, belts, spark plugs, and brake pads for each of your vehicles. These are the most common items that need replacement, and having them available ensures minimal downtime.

Best Practices: Label and organize your parts for easy access, and rotate them regularly to ensure nothing expires or becomes obsolete.

Practical Tip: Swap or barter spare parts with neighbors who have similar vehicles to expand your stockpile without additional cost.

57. Navigating Without GPS

In the event that GPS systems fail or you're in an area without reliable signals, learning to navigate using traditional methods is essential.

Map and Compass Skills

Using a map and compass is a time-tested way to navigate without relying on technology. Mastering this skill is crucial for off-grid adventurers.

Basic Compass Use

Learn how to orient your compass with a map to determine your location and navigate to your destination.

Best Practices: Practice using a map and compass regularly, even if you're familiar with the terrain. This ensures your skills are sharp when you need them.

Practical Tip: Carry a waterproof map of your area and a compass at all times, even on short journeys.

Navigating by the Stars and Landmarks

In situations where even a map and compass aren't available, natural navigation methods such as using the stars, sun, or landmarks can help you stay on course.

Using the North Star

The North Star is a reliable guide for finding direction in the northern hemisphere. By locating this star, you can determine which direction is north and orient yourself accordingly.

Best Practices: Learn to identify key constellations, like Ursa Major and Orion, that can guide you in both northern and southern hemispheres.

Practical Tip: Practice nighttime navigation with the stars regularly to build confidence in using this method.

58. Seasonal Mobility

Different seasons bring unique transportation challenges. Whether you're dealing with snow, floods, or dry conditions, it's important to have transportation methods that can adapt to seasonal changes.

Winter Mobility: Snowmobiles, Sleds, and Animal Transport

Winter presents challenges for off-grid travel, but snowmobiles, dog sleds, and even animal-powered transport like horses can provide reliable mobility during snowy months.

Maintaining Snowmobiles

Snowmobiles are excellent for covering long distances in the snow, but they require specific maintenance to ensure they're ready for winter use.

Best Practices: Before winter arrives, check the tracks, belts, and engine for wear and tear. Store your snowmobile in a sheltered area to prevent snow and ice buildup.

Practical Tip: Create a trail map of your property and surrounding areas so you can navigate efficiently during heavy snowfall.

Preparing for Flooded Areas

In regions prone to seasonal flooding, having a boat or raft ready for emergencies ensures you can navigate safely through flooded areas.

Storing Emergency Watercraft

Store your boat or raft in an easily accessible location, and ensure it's stocked with life jackets, emergency supplies, and navigation tools.

Best Practices: Regularly check your boat's buoyancy and motor functionality (if applicable) to ensure it's ready for use in an emergency.

Practical Tip: Practice loading your family and essential supplies into the boat during dry conditions to streamline the evacuation process if flooding occurs.

Module M | Community Building & Networking

Living off the grid doesn't mean living in isolation. Building a community with other off-gridders can provide a supportive network, reduce costs, and create a shared sense of purpose. This chapter explores how to form off-grid communities, build cooperative trade networks, share infrastructure, and navigate the legal considerations of communal living.

59. Forming Off-Grid Communities

Building a supportive network with fellow off-gridders enhances your experience by sharing resources, skills, and ideas. Forming off-grid communities provides not only practical benefits but also fosters friendships and collaboration.

Finding Other Off-Gridders: Where to Begin

The first step in forming an off-grid community is finding others with similar values and goals. While off-gridders may live in remote or rural areas, there are several ways to connect with like-minded individuals.

Local Networks and Gatherings

Many regions with off-grid populations host farmers' markets, community fairs, or sustainability workshops, providing excellent opportunities to meet other off-gridders.

Best Practices: Attend local events where homesteaders, farmers, or DIY enthusiasts gather. You may find like-minded individuals who are also living off-grid or are interested in doing so.

Practical Tip: Host your own off-grid meet-up at your homestead or a public space. Encourage others to share their experiences and tips, and facilitate discussions on challenges and successes.

Online Communities and Social Media

In addition to local events, there are many online forums, Facebook groups, and websites dedicated to off-grid living. These platforms allow you to connect with others, share ideas, and even organize in-person meetups.

Best Practices: Join online groups where off-gridders share their experiences, post questions, and offer advice. Look for groups specific to your region to find nearby neighbors and potential collaborators.

Practical Tip: Start an online discussion or blog about your off-grid journey, and encourage others to contribute. This can lead to connections with people who share your values and are looking to form a community.

Establishing Off-Grid Communities

Once you've connected with others, the next step is to create an off-grid community that shares resources and responsibilities. Cooperative living can reduce costs and workload while building a supportive, sustainable network.

Defining Community Goals and Values

Before forming an off-grid community, it's important to discuss and agree on the community's core goals and values. This ensures that everyone is aligned and working toward the same objectives.

Best Practices: Hold regular meetings to discuss long-term goals, such as shared food production, infrastructure projects, and sustainability initiatives. Make sure every member has a say in decisions.

Practical Tip: Create a community charter that outlines your shared vision, rules, and responsibilities. This helps clarify roles and prevents conflicts down the road.

Organizing Labor and Resources

In a cooperative off-grid community, sharing labor and resources can reduce the burden on each individual or family. This might include shared gardens, community-built infrastructure, or rotating chores.

Best Practices: Establish a system for dividing labor and responsibilities, such as rotating who tends the garden, repairs infrastructure, or manages the community's shared resources.

Practical Tip: Use a workshare model where each community member contributes a set number of hours per week toward communal tasks like food production, infrastructure maintenance, or teaching skills.

60. Mutual Aid and Barter Systems: Trading Skills, Goods, and Services

In off-grid communities, traditional currency may not always be necessary. Barter systems and mutual aid arrangements allow community members to trade goods, services, and skills without relying on money, helping everyone meet their needs while fostering a sense of trust and cooperation.

Establishing a Local Barter Network

A barter system lets individuals trade skills and resources directly, bypassing the need for money. For example, you might trade produce from your garden for help with carpentry or solar panel installation.

Setting Up a Community Barter System

Start by organizing a local barter group where each member lists the skills or goods they can offer. Encourage regular meetups where members can exchange goods and services.

Best Practices: Create a barter board (physical or online) where members post what they have to offer and what they're looking for. This system makes it easy for members to find trades and ensures that everyone's needs are met.

Practical Tip: Use a simple barter ledger to track exchanges, especially if services are offered over time, to ensure fairness and transparency.

Building a Regional Trade Network

For larger-scale bartering, consider forming a regional trade network with other off-grid communities. This can expand the range of available goods and services and strengthen connections with nearby homesteads.

Best Practices: Partner with nearby communities to exchange goods such as seeds, tools, or livestock. Establish regular regional markets where multiple communities come together to trade.

Practical Tip: Create a mutual aid calendar where community members post events like tool-sharing days, group gardening, or communal building projects, making it easier to organize collaborative efforts.

Skill Sharing: Building Knowledge Through Mutual Aid

In an off-grid community, skills like carpentry, herbal medicine, and food preservation are valuable assets that can be shared among members. Teaching and learning from each other not only builds stronger individuals but also strengthens the community as a whole.

Hosting Skill-Sharing Workshops

Encourage community members to host workshops where they teach skills like soap-making, animal husbandry, or mechanical repair. These workshops can rotate between members' homesteads.

Best Practices: Organize workshops based on the season or community needs. For example, host a canning workshop in the fall or a woodworking class in the winter.

Practical Tip: Offer a skills exchange where members can trade workshop attendance for teaching another class, creating a continuous loop of learning.

61. Shared Infrastructure, Collaborating on Essential Community Projects

One of the most significant benefits of an off-grid community is the ability to collaborate on shared infrastructure projects like water systems, solar farms, or medical aid stations. By pooling resources and labor, the community can build more extensive and durable systems than any one individual could alone.

Water Infrastructure: Shared Wells and Rainwater Harvesting

Water is essential for off-grid living, and building shared water systems can make sourcing and storing water more efficient for everyone.

Drilling and Maintaining a Shared Well

A shared well can serve multiple households in a community, reducing the individual cost of drilling and maintenance. Everyone shares the responsibility of keeping the well in good condition.

Best Practices: Assign a well maintenance team to check the water quality, monitor the pump, and handle repairs. Ensure that everyone contributes to a maintenance fund for repairs and upgrades.

Practical Tip: Rotate responsibilities for water system checks and repairs, so the load is evenly distributed across the community.

Building a Rainwater Harvesting System

For communities in areas with regular rainfall, a rainwater harvesting system can be an effective way to provide water for irrigation, washing, and even drinking with proper filtration.

Best Practices: Collaboratively build rain barrels, cisterns, and filtration systems that collect rainwater from rooftops. Distribute the water among households according to need.

Practical Tip: Set up communal water storage tanks where each household can draw from the supply, ensuring equitable access.

Solar Farms and Energy Solutions

Off-grid communities can benefit from shared solar farms and renewable energy solutions that generate power for multiple households.

Building a Shared Solar Power System

Solar farms are a great way to generate renewable energy for the entire community. By pooling resources, members can install more extensive solar systems than they could individually.

Best Practices: Designate a team to manage the installation and maintenance of the solar panels, batteries, and inverters. Allocate power based on each household's usage or needs.

Practical Tip: Set up a solar power monitoring system so that each household can track their energy use and adjust as necessary to ensure fair distribution.

Managing Communal Energy Projects

In addition to solar power, communities can also explore wind or hydroelectric power if they are in the right environment. These projects require a collective effort to build and maintain but can provide significant energy for the group.

Best Practices: Start small by installing a wind turbine or micro-hydro system, and gradually expand as the community grows. Ensure that everyone shares in both the work and the benefits of the project.

Practical Tip: Use the energy generated to power community tools or shared appliances, such as freezers or water pumps, making sure that essential infrastructure is prioritized.

62. Legal Considerations for Communal Living

When living in an off-grid community, it's essential to understand the legal aspects of shared land, property rights, and communal ownership. This section explores how to navigate the legalities of communal living and protect your community's interests.

Shared Land Ownership: Deciding How to Structure Property Rights

In off-grid communities, land ownership can be structured in several

ways, from individual ownership with shared responsibilities to communal land trusts.

Individual Ownership with Shared Responsibilities

In this model, each household owns a specific parcel of land, but community members share certain resources and responsibilities, like water systems or roads.

Best Practices: Clearly outline each household's property boundaries and responsibilities in a community agreement. Ensure that shared resources are maintained fairly.

Practical Tip: Establish a land-use committee to resolve any disputes over property boundaries or shared resource management.

Forming a Land Trust

Another option is to form a land trust, where the community collectively owns and manages the land. Each member has equal rights to the property, and decisions are made as a group.

Best Practices: Create a legal framework for the land trust, outlining each member's rights and responsibilities. Make decisions collectively to ensure that everyone has a say in how the land is used and managed.

Practical Tip: Consult with a legal professional experienced in land trusts or cooperatives to ensure that your agreement complies with local laws.

Zoning and Building Codes: Understanding Local Regulations

Even in rural areas, zoning laws and building codes may affect how your community develops its infrastructure. Understanding these regulations is crucial to avoid legal issues.

Navigating Zoning Laws

Zoning laws dictate what types of structures can be built on certain types of land and how the land can be used. Make sure your community is aware of local regulations before starting any major building projects.

Best Practices: Work with local officials to obtain the necessary permits and approvals for communal projects like wells, roads, or energy systems. Be proactive in addressing any legal concerns early on.

Practical Tip: If zoning laws are restrictive, consider applying for a variance or forming a legal cooperative to obtain special permissions for off-grid projects.

Complying with Building Codes

Building codes regulate the safety and design of structures, including homes, barns, and communal buildings. Compliance ensures that your community's buildings are safe and sustainable.

Best Practices: Designate a building committee to oversee construction projects and ensure they meet safety standards. Keep records of inspections and permits to avoid legal complications.

Practical Tip: Build using natural materials like adobe or straw bale, which may have different building code requirements than traditional materials. Consult with local experts to ensure compliance.

Module N | General Emergency Preparedness Tips

Off-grid living offers a sense of freedom, but it also comes with the responsibility of preparing for emergencies. Whether it's a natural disaster or an unexpected event, being prepared ensures your safety and survival. This chapter will cover essential tips for emergency preparedness including building a secondary shelter, planning for evacuations, and teaching children survival skills. By the end of this section, you'll feel confident in your ability to handle life off-grid even in extreme circumstances.

63. Off-Grid Shelter and Bug-Out Location

One of the key elements of emergency preparedness is having a bug-out location or secondary shelter in case your primary home becomes unsafe. This shelter should be equipped with the essentials for survival and easy to access when needed.

Choosing the Right Location for Your Bug-Out Shelter

When selecting a bug-out location, consider factors like accessibility, proximity to natural resources, and security. The location should be far enough from potential dangers but close enough for you to reach quickly in an emergency.

Proximity to Water and Resources

Your bug-out location should be near natural water sources like streams, lakes, or springs, and in an area with abundant wood or forageable plants for long-term sustainability.

Best Practices: Choose a site that has natural windbreaks, such as hills or trees, to protect your shelter from extreme weather.

Practical Tip: Map out several routes to your shelter, ensuring you have backup escape paths in case one becomes inaccessible due to flooding, fires, or other hazards.

Building a Simple and Functional Shelter

Your bug-out shelter should be designed to provide immediate protection and can be as simple or elaborate as your needs and resources allow. Common structures include log cabins, yurts, or earth shelters.

Best Practices: Build using locally sourced materials to minimize costs and environmental impact. Ensure the structure is well-insulated and can be heated or cooled with minimal resources.

Practical Tip: Use natural camouflage to help your shelter blend into the environment, making it less visible to outsiders during emergencies.

Maintaining Your Secondary Shelter

Regular maintenance of your bug-out shelter is key to ensuring it remains in good condition when you need it. Inspect it seasonally, checking for signs of wear, leaks, or damage.

Stocking Your Shelter with Supplies

Your secondary shelter should be stocked with essential supplies like non-perishable food, first aid kits, tools, and water filtration systems.

Best Practices: Rotate your stock of food and medical supplies regularly to ensure everything remains fresh and functional.

Practical Tip: Keep multiple caches of supplies in your shelter, both inside and hidden around the property, to ensure you're not left empty-handed in case of theft or disaster.

64. Tips on Life Without Essentials: Adapting to Off-Grid Challenges

Living off-grid sometimes means going without the comforts and resources of modern life. Learning how to adapt to these situations builds resilience and ensures that you can survive even when essential supplies run out.

Managing Without Power

When living off-grid, power outages or a lack of access to electricity are inevitable. Being prepared for life without power involves learning to manage daily tasks without modern appliances.

Alternative Cooking Methods

Without electricity or propane, you'll need to rely on alternative cooking methods like wood-fired stoves, rocket stoves, or solar ovens.

Best Practices: Keep a backup supply of firewood and kindling, and practice cooking meals over an open flame to hone your skills.

Practical Tip: Build a rocket stove using minimal materials. It's an efficient cooking system that can be powered by small twigs and branches.

Lighting Without Electricity

In the absence of power, you can use solar-powered lights, oil lamps, or even homemade candles to illuminate your space.

Best Practices: Keep a supply of wicks and oil on hand, and learn how to make candles from beeswax or animal fat.

Practical Tip: Install reflective surfaces around your home to maximize the light produced by your oil lamps or candles, reducing the need for multiple light sources.

Conserving Water When Supplies Run Low

When water becomes scarce, knowing how to stretch your supply can be the difference between comfort and survival.

Greywater Recycling

Use greywater systems to recycle water from sinks, showers, and washing to irrigate gardens or flush toilets.

Best Practices: Install a simple greywater filtration system to make sure the water is safe for reuse in irrigation or cleaning.

Practical Tip: Set up a rainwater harvesting system to collect and store water during wet seasons, ensuring a backup supply during droughts.

Waterless Hygiene Practices

When fresh water is scarce, you can stay clean using dry hygiene practices such as dry shampoo, baking soda, and wipes made from natural materials.

Best Practices: Create a hygiene kit that includes waterless soap, rubbing alcohol, and sanitizing wipes made from cloth that can be reused.

Practical Tip: Learn how to make your own dry shampoo from cornstarch or arrowroot powder to stay fresh without using water.

65. Evacuation Planning: Preparing for Natural Disasters and Emergencies

Even the best off-grid setup can't prevent every disaster. Planning for potential evacuations ensures you can leave safely and quickly if necessary.

Creating a Family Evacuation Plan

Evacuations can happen suddenly, so it's important to have a detailed plan that everyone in your household understands.

Identifying Emergency Escape Routes

Map out multiple escape routes from your property, considering all possible scenarios like floods, wildfires, or road blockages.

Best Practices: Identify safe locations where you can stay temporarily, such as public shelters, friends' homes, or other off-grid locations.

Practical Tip: Practice evacuation drills regularly with your family to ensure everyone knows the steps to take during an emergency.

Preparing an Evacuation Bag (Bug-Out Bag)

Each family member should have an evacuation bag packed with essentials like food, water, clothing, and first aid supplies.

Best Practices: Keep your bug-out bags lightweight but packed with enough supplies to last 72 hours. Include maps and compasses in case GPS systems fail.

Practical Tip: Store your bug-out bags in a central, easy-to-reach location so you can grab them quickly during an emergency.

Staying Informed and Communicating

During an emergency, staying informed about the situation and communicating with loved ones is crucial.

Using Radios for Communication

In an off-grid setting, radios like HAM radios or walkie-talkies provide essential communication when cell towers are down or unreliable.

Best Practices: Learn how to operate a HAM radio, and ensure you have a charged hand-crank or solar-powered radio for updates during emergencies.

Practical Tip: Set up a communication plan with neighbors or nearby off-gridders so you can relay important information during a disaster.

66. Children Preparedness: Teaching Survival Skills Early

Preparing children for emergencies is critical in off-grid living. Teaching them basic survival skills gives them confidence and ensures they know what to do if they're ever separated from adults.

Teaching Basic Survival Skills

Children should be taught the essential skills they need to stay safe in off-grid or wilderness environments.

Fire-Making and Water Filtration

Teaching children how to build a fire and filter water safely gives them the knowledge to survive in a range of scenarios.

Best Practices: Supervise children as they practice using fire-starting tools like flint and steel or matches, and show them how to use a water filter to make water safe for drinking.

Practical Tip: Set up a mini survival course where children can practice their skills in a controlled environment, learning how to build fires, find water, and stay warm.

Building Simple Shelters

Children can learn to build basic shelters like lean-tos or debris huts, using materials found in the environment.

Best Practices: Show them how to collect leaves, branches, and bark to build an effective shelter that protects them from the elements.

Practical Tip: Encourage children to sleep in their shelters during a camping trip to help them become comfortable with the experience.

Teaching Emergency Signals

Children should know how to signal for help in case they're lost or in danger.

Using Whistles and Signal Fires

Whistles are an excellent tool for children to use to signal their location. Additionally, they should know how to build a signal fire in an open area to attract attention.

Best Practices: Provide each child with a loud whistle and teach them to blow it in patterns of three, which is the universal distress signal.

Practical Tip: Teach children how to create a signal fire safely by burning green leaves or wet wood to produce smoke that can be seen from a distance.

67. Essential Wilderness Survival Skills: Staying Safe in Nature

Learning wilderness survival skills is critical in off-grid living, especially in remote areas where professional help may be unavailable for extended periods.

Fire-Making Techniques

Knowing how to start a fire is one of the most critical survival skills, providing warmth, cooking ability, and a way to signal for help.

Using Flint and Steel or a Ferro Rod

Flint and steel or ferrocerium rods are reliable fire-starting tools that work even in damp conditions.

Best Practices: Practice starting a fire with tinder, kindling, and larger wood, and learn how to keep it burning steadily.

Practical Tip: Keep waterproof matches and a ferro rod in your emergency kit, as they can be lifesavers in wet conditions.

Shelter-Building in the Wild

Building a shelter is essential for protecting yourself from the elements in wilderness survival situations.

Constructing a Lean-To Shelter

A lean-to shelter is simple to build and provides protection from wind, rain, and snow.

Best Practices: Learn to find natural materials like branches, leaves, and moss to create an insulated, wind-resistant structure.

Practical Tip: Position your shelter near a heat-reflective surface, like a rock face, to retain warmth from a nearby fire.

68. Quick Useful Knots to Learn

Knowing how to tie a variety of knots is an invaluable skill in off-grid living, from securing loads to building shelters.

The Square Knot: Securing Items Safely

A square knot is a versatile and easy-to-learn knot used for tying two ends of a rope together securely.

How to Tie a Square Knot

This knot is created by tying one end of the rope over and under the other, and then repeating the motion in reverse.

Best Practices: Use the square knot for securing bundles, tying bandages, or connecting short ropes.

Practical Tip: Practice tying and untying the square knot to ensure you can do it quickly in emergencies.

The Bowline: Creating a Secure Loop

A bowline knot forms a secure loop at the end of a rope and is ideal for rescue situations or tying animals.

How to Tie a Bowline Knot

To create a bowline, form a loop, pass the end of the rope through it, and pull tight to secure.

Best Practices: Use this knot to fasten a rope to trees, stakes, or poles when building shelters.

Practical Tip: Teach this knot to children, as it's easy to learn and extremely useful in survival situations.

The Clove Hitch:

The clove hitch is a simple knot ideal for quickly securing a rope to a post, tree, or pole.

How to Tie a Clove Hitch

Wrap the rope around the post twice, crossing over the second loop, and tuck the end under the crossing point to secure.

-Best Practices: Use the clove hitch for securing tents, tarps, or tying gear to poles.

-Practical Tip: This knot is great for temporary fastenings but should be monitored if used for longer periods.

The Taut-Line Hitch: Adjustable Tension Knot

A taut-line hitch is perfect for creating an adjustable loop that can slide to tighten or loosen tension, making it ideal for securing tents or tarps.

How to Tie a Taut-Line Hitch

Wrap the rope twice around the standing line, then make a final loop on the outside before pulling tight to secure.

Best Practices: Use this knot for adjustable tension in guy lines or securing loads that may shift.Practical Tip: Practice tightening and loosening the knot to ensure it stays secure under pressure.

The Figure-Eight Knot: Strong and Secure Stopper Knot

A figure-eight knot creates a strong stopper at the end of a rope, preventing it from slipping through loops or holes.

How to Tie a Figure-Eight Knot

Form a loop in the rope, then pass the end around the standing line and back through the loop to form a figure-eight shape.

Best Practices: Use the figure-eight knot to prevent ropes from fraying or slipping through pulleys.

Practical Tip: This knot is often used in climbing and rescue operations due to its strength and simplicity.

The Sheet Bend: Joining Two Ropes

The sheet bend is ideal for tying two ropes of different sizes or materials together securely.

How to Tie a Sheet Bend

Pass the thicker rope in a loop, then bring the thinner rope through the loop, around the back, and tuck it under itself.

Best Practices: Use the sheet bend when you need to connect ropes of different diameters, especially in emergency situations.

Practical Tip: Always finish the knot with a double sheet bend for added security when working with slippery or wet ropes.

The Trucker's Hitch: Creating Mechanical Advantage

The trucker's hitch is perfect for creating tension to secure heavy loads or tighten down tarps.

How to Tie a Trucker's Hitch

Create a loop in the middle of the rope, pass the end through a fixed point, and pull it back through the loop for mechanical advantage before tying off.

Best Practices: Use the trucker's hitch for tying down cargo, securing tents, or anchoring loads tightly.

Practical Tip: Practice tying this knot to quickly adjust tension when securing large or heavy items.

Conclusion

Living off the grid is more than just a lifestyle—it's a deliberate choice to step away from modern conveniences and embrace self-sufficiency, independence, and resilience. As you've seen throughout this guide, the journey to off-grid living involves numerous skills, careful planning, and a mindset that prioritizes adaptability, resourcefulness, and environmental stewardship. Whether you're just beginning your off-grid adventure or have years of experience under your belt, the principles and projects detailed in this guide are meant to equip you with practical knowledge to thrive in any situation.

The Importance of Planning and Preparation

Planning is the cornerstone of any successful off-grid project. From selecting the right location to ensuring you have the infrastructure to support your family's needs, the first steps you take will define the success of your off-grid lifestyle. As outlined in the early chapters, choosing the right location involves balancing natural resources, legal considerations, and community dynamics. Every decision, from how to build your home to sourcing your water, will shape your ability to live sustainably and independently.

By carefully mapping out your homestead's layout, building with local and sustainable materials, and accounting for factors like climate and natural disasters, you're preparing yourself for a long-term lifestyle that prioritizes self-sufficiency. The same level of meticulous preparation applies to ensuring your home is energy-efficient, properly insulated, and oriented to maximize sunlight and protection from the elements.

Water Independence: The Foundation of Sustainability

Water is life, and securing reliable water systems is essential for off-grid survival. Whether you're harvesting rainwater, digging a well, or utilizing natural bodies of water, the process requires both planning and safeguards. This book has covered water collection, purification, distribution, and waste management systems to help you stay water-secure regardless of your location.

The sourcing of water is the first challenge, but learning to purify and distribute that water sustainably is just as critical. The techniques we've discussed, from slow sand filtration to UV purification, ensure you have access to safe drinking water. Proper wastewater management techniques, like greywater recycling and septic system maintenance, complete the cycle of water sustainability.

Energy Independence: Powering Your Off-Grid Life

Reliable energy is a central element of off-grid living, and your ability to generate, store, and manage power affects every part of your homestead. This book has emphasized a wide range of energy production methods—from solar panels to wind turbines and hydroelectric systems—allowing you to tailor your setup to the resources available at your site.

Hybrid energy production methods and backup generators ensure you're prepared for power fluctuations, and battery banks provide energy storage for times when production is low. By combining these systems with energy-efficient appliances and strategies for heat retention, you can minimize waste and maximize the longevity of your resources.

Waste Management: Handling Off-Grid Challenges

One of the unique challenges of off-grid living is managing waste sustainably. From building composting toilets to developing greywater systems, you've learned how to handle human waste, minimize environmental impact, and turn waste into valuable resources like compost or biogas.

By focusing on waste minimization strategies, reusing materials, and safely handling hazardous waste, you've developed a holistic approach to waste management that ensures your homestead remains both clean and environmentally responsible.

Sustainable Food Production and Procurement

Food is another pillar of self-sufficiency. Whether you're growing your own food, raising animals, or foraging in the wild, learning to provide for yourself without relying on modern supply chains is liberating. This book has walked you through methods like gardening, permaculture, and hydroponics to create a thriving food system on your property.

Equally important is the skill of preserving food through canning, pickling, dehydrating, and smoking. These methods allow you to store excess produce for the winter months, creating a buffer against seasonal shortages or unexpected events. Beyond growing and preserving food, foraging, fishing, and hunting offer alternative methods for feeding your family off the land.

No Grid Cooking, Heating, Cooling, and Lighting Systems

As we've covered, living without modern infrastructure means developing off-grid cooking, heating, cooling, and lighting systems. By embracing traditional methods, like rocket stoves, wood-fired ovens, and solar ovens, you can prepare meals without relying on fossil fuels. For heating and cooling, using natural insulation, passive solar design, and efficient wood stoves allows you to maintain a comfortable living environment year-round.

Lighting systems, from solar-powered lights to oil lamps, provide alternatives to electric lighting that are both functional and sustainable. Learning to make candles or use natural light sources ensures you're never left in the dark, even in extended power outages.

Health and Wellbeing: Staying Strong Off-Grid

Physical and mental wellbeing is critical in an off-grid lifestyle, where self-reliance extends to health and medical care. This book has emphasized the importance of healthy living through diet, exercise, and mental clarity. With access to natural remedies, herbal medicine, and first aid skills, you're equipped to handle most medical issues that arise. By stocking basic medications and learning natural alternatives, you can avoid dependency on external healthcare systems for minor issues.

Sanitation is another crucial element of staying healthy off-grid. Hygiene best practices, especially when running water isn't always available, ensure you and your family remain disease-free and comfortable.

Security and Fortification: Protecting Your Homestead

Security plays an essential role in off-grid living, where outside help may not be immediately available. We've covered how to assess vulnerabilities, fortify your home, and create safe rooms or shelters in case of emergencies.

From installing security systems to utilizing firearms, guard animals, or non-lethal traps, off-grid security strategies ensure your property is safe from intruders and wildlife alike. Understanding legal considerations and maintaining firearm safety are key to developing a balanced and secure homestead.

Technology, Communication, and Financial Independence

Technology, though limited, still plays an essential role in off-grid living. By leveraging low-tech tools like HAM radios, walkie-talkies, and satellite phones, you can maintain communication with the outside world when needed. Likewise, understanding financial preparedness—from budgeting for an off-grid lifestyle to developing monetization opportunities—ensures your long-term financial sustainability.

Living off-grid often means finding creative ways to earn income, whether through consulting, selling handmade goods, or offering training and skills-sharing to others seeking to embrace self-sufficient living.

Education and Skill Acquisition: Lifelong Learning Off-Grid

Education is an ongoing journey, particularly when living off-grid. You've learned that acquiring and sharing skills within a community is essential for both personal growth and community resilience. From practical DIY skills like carpentry and blacksmithing to teaching children critical survival skills, off-grid education fosters a culture of independence and ingenuity.

Whether you're homeschooling children, learning through online resources, or teaching others how to master essential skills, lifelong education ensures everyone in the community thrives.

Community and Mental Wellbeing

Living off-grid doesn't have to mean living alone. By building communities and engaging in mutual aid networks, you're creating a safety net of support, collaboration, and shared knowledge. The emotional resilience required for off-grid living can be sustained through mindfulness practices, meditation, and staying active through hobbies like gardening, woodworking, or storytelling.

Keeping social connections alive, whether through group activities, board games, or community-building efforts, helps prevent isolation and ensures a sense of belonging in your off-grid journey.

General Emergency Preparedness

No off-grid living manual would be complete without emphasizing emergency preparedness. Whether it's securing a bug-out location, teaching children survival skills, or practicing evacuation plans, emergency preparedness ensures you're ready for any situation.

From learning to tie essential knots to mastering wilderness survival skills, being ready for the unexpected is at the core of off-grid living. Planning for life without essentials, managing limited resources, and maintaining self-reliance during emergencies are what define the off-grid experience.

Final Thoughts: The Journey of Off-Grid Living

As you embark or continue your off-grid journey, remember that resilience, adaptability, and a willingness to learn are your greatest assets. This lifestyle isn't just about surviving—it's about thriving in harmony with nature, building meaningful connections, and fostering a deep sense of personal independence.

Through careful planning, acquiring essential skills, and maintaining a balance between self-reliance and community collaboration, you are fully equipped to lead a fulfilling, sustainable, and empowered off-grid life. Whether you're working with the land, building your own home, or living independently from modern conveniences, the knowledge and skills shared in this guide will serve as your roadmap to a self-sufficient and resilient future.

Thank You

Creating this guide has been a journey driven by dedication and a passion for providing you with the essential knowledge and hands-on skills needed to thrive in a world where self-reliance is key. Your support plays a crucial role in motivating us to continue empowering others to embrace off-grid living and survival. Thank you for being an important part of this mission!

We'd love to hear your thoughts! Your feedback is incredibly valuable to us and to others who are seeking practical ways to live off the grid. Sharing your thoughts is quick and easy, and only takes a minute. You can choose how you'd like to share your experience:

Option A: Record a short video sharing your favorite parts or how you're using what you've learned.

Option B: If you're camera-shy, no problem! You can take a few pictures of the book or simply write a few sentences. Your input not only acknowledges our work but also helps others discover this valuable resource.

You can share your feedback by clicking the provided link or scanning the QR code below. While it's entirely optional, your feedback means the world to us.

Scan the QR code below to share your excitement!

Retrieve Your Bonus Content

Thank you for your support! To access the bonus content, simply visit the link or scan the QR code below.

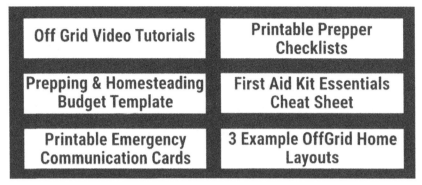

Or proceed to this link: rebrand.ly/ng-extra

References

- "Hydroponics", Springer Science and Business Media LLC, 2024
- 247broadstreet.com
- 8msolar.com
- acuacultura.org
- aiswindows.com
- allaboutchemistry.net
- alternet.org
- Amit Kumar Pandey, Sanjeevikumar Padmanaban, Suman Lata Tripathi, Vivek Patel, Vikas Patel. " Microgrid - Design, Optimization, and Applications", CRC Press, 2024
- appelheat.com
- architecturecourses.org
- besttoiletbuy.com
- bidhya.com
- blog.ecoflow.com
- botanikks.com
- brightideas.houstontx.gov
- buildingmep.com
- chicagoartmagazine.com
- chickenkeepingsecrets.com
- Chris Park. "The Environment - Principles and Applications", Routledge, 2019
- Christopher J. Rhodes. "Feeding and Healing the World: Through Regenerative Agriculture and Permaculture", benchmarkhydroponics.com.au
- cpss.net
- D. Yogi Goswami, Frank Kreith. "Handbook of Energy Efficiency and Renewable Energy", CRC Press, 2019
- energy5.com
- energymatters.com.au
- engineeringsadvice.com
- entekhvac.com
- Ernest R. Vieira. "Elementary Food Science", Springer Science and Business Media LLC, 1996
- fark.com
- fastercapital.com
- foreignpolicy-infocus.org
- Frank R. Spellman. "Handbook for Waterworks Operator Certification - Fundamental Level, Volume I", CRC Press, 2019
- friendlyaquaponics.com145
- Guodao Zhang, Haijun Zhou, Yisu Ge, Sharafzher M. Magabled et al. "Enhancing ongrid renewable energy systems: Optimal configuration and diverse design strategies", Renewable Energy, 2024
- homecompostingsolutions.com
- hydrogardengeek.com
- hydroone.com
- Ibrahim Dincer, Dogan Erdemir. "Chapter 1 Solar Energy Systems", Springer Science and Business Media LLC, 2024

- iihtchampa.in
- kaiweets.com
- Kassian T.T. Amesho, E.I. Edoun, Timoteus Kadhila, Sumarlin Shangdiar et al. "Technologies to convert waste to bio-oil, biochar, and biogas", Elsevier BV, 2024
- Long Shi, Haihua Zhang. "Solar Chimney Applications in Buildings", Springer Science and Business Media LLC, 2024
- lukasqjybp.post-blogs.com
- Maulin P. Shah, Pardeep Kaur. "Biomass Energy for Sustainable Development", CRC Press, 2024
- mindcull.com
- nimbusfacility.com
- picturethisai.com
- pluginhighway.ca
- Randall Thomas. "Environmental Design 2e", Routledge, 2019
- Seong-Ryong Ryu, Kyu-Nam Rhee, Myoung-Souk Yeo, Kwang-Woo Kim. "Strategies for flow rate balancing in radiant floor heating systems", Building Research & Information, 2008
- solarplanet.uk
- Steven Szokolay. "Introduction to Architectural Science", Routledge, 2019
- sunshineonmyshoulder.com
- utilitiesone.com
- viveirosinsurance.com
- wastewatersolutions.us
- whereisthenorth.com
- wikifarmer.com
- worldhealth.net83
- Yahya Alassaf. "Comprehensive Review of the Advancements, Benefits, Challenges, and Design Integration of Energy-Efficient Materials for Sustainable Buildings", Buildings, 2024
- Yashar Aryanfar, Mamdouh El Haj Assad, Jorge Luis García Alcaraz, Julio Blanco Fernandez et al. "A thorough review of PV performance, influencing factors, and mitigation strategies; advancements in solar PV systems", Elsevier BV, 2024

Made in the USA
Las Vegas, NV
26 November 2024